Enduring Erosions

Enduring Erosions

Environmental Displacement and
Relocation on India's Sinking Coasts

Arne Harms

HAWAI

University of Hawai'i Press
Honolulu

This book is freely available in an open access edition thanks to funding from the Max Planck Institute for Social Anthropology Open Access Fund.

Paperback edition 2025
Printed in the United States of America
First printed, 2024

Library of Congress Cataloging-in-Publication Data
Names: Harms, Arne, author.
Title: Enduring erosions : environmental displacement and relocation on
  India's sinking coasts / Arne Harms.
Description: Honolulu : University of Hawai'i Press, [2024] | Includes
  bibliographical references and index.
Identifiers: LCCN 2024002849 (print) | LCCN 2024002850 (ebook) | ISBN
  9780824897536 (hardback) | ISBN
  9798880701759 (epub) | ISBN 9780824898755 (kindle edition) | ISBN
  9798880701742 (pdf)
Subjects: LCSH: Environmental refugees—Sundarbans (Bangladesh and India) |
  Forced migration—Environmental aspects—Sundarbans (Bangladesh and
  India) | Coast changes—Sundarbans (Bangladesh and India) | Climatic
  changes—Social aspects—Sundarbans (Bangladesh and India) | Sundarbans
  (Bangladesh and India) —Environmental conditions.
Classification: LCC HV640.5.I4 H38 2024 (print) | LCC HV640.5.I4 (ebook)
  | DDC 362.87095492/5—dc23/eng/20240516
LC record available at https://lccn.loc.gov/2024002849
LC ebook record available at https://lccn.loc.gov/2024002850
  ISBN 9780824898885 (paperback)

Cover: Erosion on high-land shores, Ghoramara Island. Photograph by author.

University of Hawai'i Press books are printed on acid-free paper and meet the guidelines for permanence and durability of the Council on Library Resources.

Contents

Acknowledgments / vii

A Note on Transliteration / ix

Chapter 1. Introduction: Relearning the Coast / 1

Chapter 2. A Postcolonial Archipelago / 34

Chapter 3. Hidden in Plain Sight: Coastal Erosion as Distributed Disaster / 44

Chapter 4. Seeking Shelter in the Mouth of the River / 86

Chapter 5. Resettlement by Other Means / 116

Chapter 6. Keeping Dry, Staying Afloat: The Politics of Coastal Protection / 144

Chapter 7. Of Blame and Protection / 172

Chapter 8. Conclusion: Mosaics and Futures / 198

References / 207

Index / 235

Acknowledgments

I LIKE to think how a research project, eventually informing a book like this, comes to be influenced by people, relations, and institutions rolling in and out like the tides. I was fortunate enough to benefit from the support, vision, hospitality, openness, and input of many people and institutions on the way, and from the relations they sustained. Like the tides, they have swept in at different moments, weaving themselves into the fabric of this book in different ways; and like currents, they have pushed me in all kinds of directions.

I owe my greatest thanks, and extend my solidarities, to people braving eroding lands, shrinking islands, and hungry tides on the western perimeters of India's Sundarbans. I deeply appreciate the time I was offered, and the kindness with which so many islanders opened their homes and hearts for my silly inquiries into what was so obvious, yet so hard to grasp. I owe so much to all of you who sat down or walked with me, who shared a cup of tea, insights, and laughter, who took a break from work to engage in this kind of exercise, who answered my nosey, sometimes clumsy inquiries, frequently with witty return questions. I enjoyed learning together. And I hope that some of you would recognize yourself in the stories and doings and relations I tell of, even if your names have been altered to give you privacy and I decided to write in this global language, instead of the beautiful vernacular, the Bengali from those coasts, you would find more accessible (if it were not for the jargon). I hope most of you still would understand the choice of language, as you did, in order to reach wider audiences.

Speaking of relations and intimacies, words fail to express the gratitude I feel toward my family and loved ones back home for all the support, care, and inspiration they extended at different stages. Sadly, my father wouldn't live to see this book go into print; my mum, however, does. Thank you both for your belief in my scribblings. I dedicate the book to you, and to my son, Lele, who gives me so much hope.

The research that went into this book would not have been possible without the mentorship provided, doors opened, and connections made by generous people all along the way. Among these are Ute Luig, Ranabir Sammadar, Joachim Schmerbeck, Mr. Giri, Anu Jalais, Partha Kayal, Sunando Bandyopadhyay, Sibdas Ghosh and the Humboldt Club, Uwe Lübken, Rebecca Hofmann, Martin Sökefeld, Silja Klepp, Kenneth Bo Nielsen, Sraman Mukherjee, Ursula Rao, Jason Cons, and Lukas Ley.

I owe special thanks to my field assistant and friend Probir Neogi, and to Uthio Mukherjee and everyone else at Omkarnath Nibas for providing shelter by the shore.

My thinking on and with eroding shores has benefitted tremendously from navigating the postdoc research landscape alongside so many bright, welcoming, and enthusiastic colleagues between Berlin, Munich, Nalanda, Leipzig, and Halle. Learning together with you on how to do, and push and tweak, this thing called anthropology never ceases to amaze me.

I have had the honor of presenting parts of the arguments developed here at conferences, workshops, and colloquia at Free University of Berlin, City of Rotterdam, University of Bremen, Rachel Carson Center Munich, Ludwig Maximillian University of Munich, Institute for Advanced Studies in the Humanities Essen, Quaid-i-Azam University Islamabad, Royal Anthropological Institute London, Durham University, University of Leipzig, and Leibnitz-Zentrum Moderner Orient Berlin. I am thankful for the impulses and critical feedback I received from the diverse, sympathetic audiences.

Thank you also to University of Hawai'i Press's anonymous reviewers for their extremely helpful critique, and to Stephanie Chun for the smooth and supportive handling of the publication process.

I am very grateful to Gregory Ward, Adriana Cloud, and Jovan Maud for straightening out my language. And to Jana Kreisl for the fantastic map.

Finally, this book would not have been possible without the generous funding of research I received from the German Academic Exchange Service, the Hans-Böckler-Foundation, the German Research Council, and the Max Planck Society.

A Note on Transliteration

THIS BOOK follows a simplified transliteration of Bengali and Sanskrit terms. Diacritic signs are omitted. Ensuing ambiguities are counterbalanced, I believe, by the convenient cross-referencing enabled by these simplifications.

Chapter 1
Introduction

Relearning the Coast

A
s the cycle rikshaw rattles along, the colors change. Above us, the sky remains blue, but the soil darkens, and the flora grows increasingly pale. We are drawing closer to the edge of the island. Soon the breeze coming in from the Bay of Bengal will offer some relief. But for now, the driver's shirt is soaking with sweat as he struggles to navigate the bumpy and pothole-laden brick road. Barren fields and thatched huts stretch across both sides of the elevated road, where children play and women tend to household tasks.

Above the creaking of the pedals, I strike up a conversation with Ganesh, the driver.[1] Both of us haven't been here in a while. He plies this road infrequently. These days, he is more likely to be ferrying pilgrims between bus stop, temple, and beach in nearby Gangasagar, where Hindus from across India gather to worship the river Ganga as she empties into the sea.[2] He knows the place where we are heading very well; friends and distant relatives call it home in the provisional ways any place can be called home on the edge of this morphing island. We are heading to Botkhali, a village sitting in the southeastern extreme of the island, flanked by the estuary on one side and the open sea on the other (see map 1.1). The village has gained notoriety across the region as one of the places most severely affected by coastal erosion. It is known for frequent embankment collapses and severe flooding, for the displacement of almost all its former residents, and for the gradual dismantling and liquification of the village itself by brackish waters. Engineers have visited in order to

---

1. All of my interlocutors have been anonymized.
2. Throughout the book, I use the spelling "Ganga" for the mighty river that is referred to in parts of the literature and maps as "Ganges." The name "Ganga" not only stays close to how Bengali-speaking islanders, or the visitors speaking one of the many North, East, or West Indian languages, address the waters, it also signposts the intricate enmeshment of divinity and materiality (see chapter 7).

Map 1.1. Map of Sagar. Drawing by Jana Kreisl. Reproduced by permission.

make assessments and formulate master plans, and journalists have dropped by, filling notebooks and camera memory with survivor stories. Islanders refer to the village as a lost cause, as the latest incarnation of a tragic process of retreat. To them, it has become a symbol of the losses wrought by an encroaching sea that has long troubled this island. To many, the village is an emblem of the maelstrom the island is in as a whole: a delicate piece of flat land sandwiched between morphing rivers and a rising sea; a frail and battered home far off the centers of the new India; a heavily populated island threatened by submergence; a bastion saving its hinterland from destruction precisely as it crumbles. Partly drawing on such assessments, journalist or humanitarian renditions insert the village into global debates as one of a string of places rendering the Sundarbans a focal point of climate change.

As we enter the last village before Botkhali, a busy market scene suddenly tugs us away from reflecting on any changes in the landscape. Surrounding a renowned temple is an eclectic array of shops and tea stalls. We stop. The quotidian bustle disguises the fact that this is the end of the road. From here on, vehicles are of no use. With the road being washed away and the former village in ruins, any further movement is to be done on foot. After some meeting and greeting, I turn to the path that leads to Botkhali. The change grips me again with full force. The soil here is even darker, most trees have been stripped of foliage, alluding to a winter that does not exist in these parts. The ponds have turned into lifeless black puddles. The broken road is now lined by makeshift huts, housing recently displaced villagers trying to settle as far as possible from the shore, yet within the parameters of their village. As life spills in and out of these tiny homes, as the tone changes all around, I can't help but think I'm entering a war zone.

A short while later, I reach the former center of the settlement. The sea has encroached further, eating deep into the village and flooding whatever is left with every spring tide. The outer embankment has been moved inland since my last visit. But with more village lands abandoned to the sea for good, it has so far done little to quell the push of the water. Outside of the perimeter, faint remnants of houses can be discerned: heightened foundations here, tree stumps there. All around it, watery expanse. The embankment, too, is in shatters. Again. Today, its remainders are little more than a slightly elevated line of muddy ground. The water rolls freely over it with every spring tide, gradually swamping what lies behind, pawing at the remaining buildings dotted across the area.

Gone are all of the makeshift huts where I visited friends last time. With the embankment rebuilt farther inland and then razed again in ensuing monsoons, they had packed up their belongings once again, moving even farther inland. Some moved out, others now stay at their relatives' or at the road that I had just passed.

Standing here, orienting myself in this new Botkhali, I realize that Ganesh has followed me. He has changed his mind, skipping tea and the chance to catch up with acquaintances on the market square, where the rikshaw is parked, and joins me. As we begin walking around, he explains that he wants to see the latest developments. Obviously, he is just as curious as I am to see the changes the last year has brought about.

Ganesh has taken me here all the way from Gangasagar, where his family has been living for more than twenty years now, a place that I called home during my research. Little known to most visitors, the pilgrimage destination of Gangasagar features a resettlement colony, originally earmarked for the relocation of people uprooted by coastal erosion. Ganesh's family—like over a hundred others—settled years ago in what they call Gangasagar Colony, or simply Colony Para, after enduring repeat displacement on the neighboring islets of Ghoramara and Lohachara (see map 1.1). First the waters took their land and home on Lohachara, and, when the islet was abandoned as a whole, they shifted to Ghoramara, which continues to straddle currents and waves. Luckily, they were included into a list of beneficiaries of land distribution in one of the resettlement colonies opened up on Sagar. Settling in Colony Para, they were escaping what I call circuits of displacement and emplacement that have shaped their lives for decades. I turn to these pasts in detail in chapter 4. Suffice to say for now that this background helps explain Ganesh's curiosity to see Botkhali and take in the most recent changes. To him it was like revisiting an already distant past, a threat that was once looming over him, as it does now over much of the shore. Our roaming through the ruins encapsulates quite well the temporal knots and mobility-induced geographical entanglements I address in this book: moments where individual pasts continue to haunt the present and globally anticipated sea level–rise futures are not only a present predicament but also already a thing of the past, where places are woven together by trajectories of movement involving people, land and water, culminating in displacement, disappearance, and reappearance elsewhere. It highlights the complicity of me as a researcher in this, activating pasts by revisiting them mostly through conversation. Wherever possible, as long as ruins were still discernable, I would strive to visit them, to immerse myself into the materiality of erosion. The places I visit and revisit in this book, then, are woven together perhaps less by chance and arbitrary decisions that sustain ethnographic fieldwork than by personal relations, narratives, and embodied practices marked by the ripples of environmental degradations.

Braving the sun, and sliding across the muddy terrain, Ganesh and I begin relearning this coast. As we roam around, we meet a villager I had only briefly met during previous visits who is now busy tearing down the roof of his former home. The house is in ruins. It collapsed partly, he says, after the incoming tides had teased its foundation for years. His family moved out, inland, he says, and all there is left for him to do is check which parts can be useful to build back elsewhere a house for his family.

Behind the house, a crew of workers wades through the mud. They dig out poles that line the land-facing shore of a narrow canal. This used to be the embankment, they tell me. Built early last year and devastated by the sea during the monsoons a few months later. With new rains approaching soon, they are employed by the state to rebuild this outer line of defense. Over the course of the next few weeks, they will repurpose the poles collected today for a wooden frame to be sunk into a new embankment, built like all preceding ones mainly from salty mud. Standing here, all of us are well aware that a few months later, at the end of the rains, nothing more will be left from their efforts than what they now have at their hands: bare poles ready to be reused for yet another embankment and yet another season.

Watching them, I realize that the school building is beyond repair, too. Between my first visit to Botkhali and today, the building had become increasingly exposed, shown cracks, fallen into disrepair, and now is a ruin. One of the very few concrete multistoried buildings in the area, it had also doubled up as a public shelter during storms and surges. But as the erosion continued unabated, fissures appeared, and a part of it collapsed. Classes are suspended, my interlocutor tells me, and they now seek shelter in other buildings when storms come.

Talking to villagers and walking on the ground, we learn about the consequences of coastal erosion. We hear, and see, about the progress made by the sea and about, so far rather futile, efforts to subdue the liquification of land by these marginalized islanders. In a sense, we confirm changes, evaluating the present against past visits, knowing very well that these are mere snapshots. We relearn the coast as we realize that all this will be undone in due course. All of what we see now and the ground we walk upon will likely be gone very soon. What we do, in other words, is trace a landscape in flux. To be sure, all kinds of places are continuously transformed at the hand of diverse classes of actors, but few with such velocity. Gnawed at and rolled over by an unruly river and a rising sea, this is a place enfolded in the drawn-out process of coastal erosion.

As we move on, a widow appears in the distance, rushing with empty containers to the next hand pump. She doesn't want to talk. Perhaps she is wary of inquisitive visitors. Or she is simply busy arranging drinking water, which is ironically difficult to come by, amid the abundance of fluids in this zone of embankment failure.

On that day, Ganesh and I were drawn together in being curious about, as well as petrified by, the ability of these powerful salty waters to swallow up entire villages dotted along this densely populated shoreline. But these are, of course, never purely natural phenomena. That water and this mud are, as I show in this book, decidedly unnatural. Their specific features are subjected to human, more-than-human and spiritual entities, and technical interventions. The unnatural state of these frail and slowly disappearing shores comes to the fore, as the subsequent chapters will illustrate, in laments about faulty politicians, state neglect, and economic marginalization. It also manifests in the cautious and reserved ways of addressing what Hindus see as the divine waters engulfing the land. Stopping short of attributions for the moment, Ganesh and I were doing what has become commonplace for people along these coasts. The population must relearn mutating landscapes and negotiate conjecture about imminent threats and opportunities. We were getting attuned to coastal erosion.

The presence of destruction and displacement is writ large here. There was little room for doubt that land would continue to disappear, and it seemed obvious that the village's very abandonment was near. Against this backdrop, getting attuned was not a purely academic or nostalgic exercise. Villagers constantly attune themselves, as I will show, in order to make informed decisions about when to abandon their homes or where to relocate. On another level, getting attuned is required in order to account for the ways in which the villagers of Botkhali, and others in similar predicaments across the Sundarbans, are coping with this situation. To account for lives and landscapes implicated in the drawn-out processes of coastal erosion calls for following trajectories of mobility and displacement, of stories and relations through which people navigate or make sense of what is happening, and for tracing practices geared at surviving at all odds. Seeing this agenda through relates accounts of enduring erosions to the quest of understanding everyday life amid marginalization and planetary injury.

Erosions, Absences, Loss

This book traces what it means to live on a landscape implicated in the slowly unfolding processes of coastal erosion. In young delta landscapes at the interface of older continental landmasses and oceans, the twin processes of erosion and accretion are the norm. Tides or floods washing over lands leave sediments behind, thus raising the land over the course of years or centuries. Similarly, sediments sink whenever the flow of the water carrying them abruptly reduces its speed, which happens, for instance, when the river meets the sea. Sinking sediments may amass and rise into accretions, perhaps washed away again very soon, perhaps rising into sandbars or islands.

When accretion outweighs erosion and the delta effectively grows land into a watery expense or above the high tide line, it is a cause for celebration, the solid rescued from the amphibious much to the liking of terrestrial forms of governance (D'Souza 2009; Elden 2007; Lemke 2015). Erosion captures the darker side of this process. Sediments get washed out, sandbars disappear, and islands may vanish. Rivers give and take, as my interlocutors say. At its extreme, erosion punctuates areas and lives as water nibbles at the landmass, unsettling chunks or collapsing embankments.

Erosion may thus be understood as a process that is inverting the hospitability of places, turning lush and productive homes into—what to most eyes would be—amorphous expanses of brackish waters. What I try to capture in this book, hence, is not simply the loss of place, but the long processes of seeing it fade out and turn increasingly thinner, increasingly lifeless. I turn to lives lived amid the slow disappearance of the ground beneath one's feet, and to the ways people resuscitate life in terms of struggling with brackish waters, of moving and settling elsewhere, of crafting pasts in order to come to terms with what it was they went through.

I am interested, therefore, in the experience of erosion as a specific form of demise and ruination that sends all kinds of ripple effects. I explore its unraveling of promises, materialities and, partly, socialities. And I explore the emergence of new socialities fostered in the enduring presence of coastal erosion: those that emerge amid the ghostly remains of formerly lush landscapes, and those that emerge elsewhere, in places removed but always still within the long shadow cast be the experience of coastal erosion and displacement.

Coastal erosion has, as I demonstrate, various temporal dimensions. It consists of drawn-out processes that frequently culminate in catastrophic instances. Liquified pasts haunt the present. Living amid eroding, lique-fying, or sinking lands means straddling an uncertain present and look-ing anxiously into what is a mostly bleak future. Erosion is nowhere only local it comes imbricated into, and informed by, wider socio-ecological political-economical processes. It is a specific mode of the social and the material being entangled in decay.

In this view, *Enduring Erosions* contributes to the growing schol-arship on life amid planetary injury (e.g., Tsing et al. 2017; Bellacasa 2017; Kawa 2016), this time by turning to everyday life and narratives along the edges of some of the fastest-shrinking settled islands along Asia's shores. I follow individual trajectories, practices, relations, and narrations on the watery edge in order to account for the ripples sent by environ-mental degradations, as well as to situate these very degradations across scales and across classes of actors involved in localized transformations. Through the prism of coastal erosion, I explore water and sediment flows as unnatural environments combining anthropogenic interventions—some nearby and recent, some removed in time and space—with vivid divinities and the characteristic unruliness of the matter water is. In short, I read coastal erosion on these shores as an articulation of the Anthro-pocene, and seek to uncover how life is being lived in its looming presence.

The concept of erosion carries myriad meanings. Contemporary social commentators frequently decry the erosion of values, for conser-vatives, or the erosion of, say, the welfare state, for those on the Left. Ero-sion threatens something crucial in a manner that is both pervasive and barely visible. Its outcomes are problematic, and—crucially—preventing it is almost always challenging. In other words, erosions appear to threaten the underpinnings of the everyday, the consequence of overarching forces working at a slow speed. In erosion, temporal modalities are knotted. Invisible at a given moment, their workings become evident only across a longer time span. Likewise, upending or undoing erosions takes time. There are no immediate solutions, and no easy fixes. A lack of consensus on how to undo erosion is then often paired with what I call a deaden-ing certainty about what is going to disappear.

On a different level, the trouble with erosion is not one of excess or disruption but rather of making something glide out of sight and disappear. As Rachel Carson (1962) famously noted, it is hard to notice the specific silence sustained by the absence of birdsong and the erosion of birds' habitats. It is therefore challenging to assess waterscapes in terms of the land they were and could become all over again.

Scholarship in environmental studies considers several articulations and contexts of erosion. Soil erosion—the depletion of topsoil aggravated by intensive farming or desertification—is the subject of debates that turned formative for the field of political ecology (Blaikie and Brookfield 1987). Furthermore, riverbank erosion is the subject of a strong body of interdisciplinary scholarship (e.g., Hutton and Haque 2004; Rahim, Mukhopadhyay, and Sarkar 2008). Both classes of phenomena—the loss of topsoil and the morphing of banks—are intimately related to one another via the flows of sediments. Washed out in one place, sediments find their way into rivers before being dropped on a massive scale as the river draws near the sea. Novel land formations bloom, only to become subject to coastal erosion by the combined forces of tides, currents, and waves that attack soft soil. Such processes are, as noted, the delta normal. The combination of large amounts of sediments and mild wave activity allows some deltas to grow very large; others are kept comparatively small by coastal flows.

This "normal" turns problematic once erosion outweighs accretion in landscapes settled by societies preferring land-based lives. And it may turn disastrous once people have to endure them within the double constrains of political economy and ecology.

Between the River and the Deep Blue Sea

This book turns to lifeworlds on a small group of coastal islands all of which are, to varying degrees, in rough waters. Situated at the southwestern edge of the active Ganges delta, at the sea-facing perimeters of an area known as the Sundarbans, these islands are in close vicinity to one another. They are drawn together, as my chapters demonstrate, by shared histories and ties of kin or friendship, situated within the same administrative units and in roughly the same pathways of storms, currents, and tides. Yet the islands differ, as do villages on these islands, and the contrasts documented in this book will help to elucidate the political ecology of loss and adaptation.

I call the islands I focus on in this book the Sagar Islands, consisting of Sagar Island and the much smaller islands of Ghoramara and Lohachara. Vernacular Bengali features a similar denominator (Sagar

Dvip, Sagar Dviper Kache). This reflects that the latter two islets grew out of the former, as the historical and geographical account validates. More importantly, Sagar will be used as a common denominator due to being by far the biggest, most populous, and best known of these islands.

Sagar is home to a population of some 120,000 inhabitants spread across hundreds of hamlets with one urban center. It benefits from two aligned development trajectories. The most recent one targets the development of its southern shore as a beach tourist destination alongside Bakkhali and Digha. The other trajectory mobilizes the island's key role within Hindu religions as a site of mythical occurrences that led to the descending of the Ganges from heaven down to earth. Not far from Botkhali—the ruined village I am constantly relearning—stands the pilgrimage complex of Gangasagar, home to an annual pilgrimage festival that attracts several hundred thousand visitors every year. Besides catering to the masses arriving here every mid-January, rendering Gangasagar Mela India's second-largest religious congregation, the pilgrimage centers also accommodate a continuous stream of pilgrims arriving from across North India to offer prayers to the divine river. Yet many of the islanders perceive this holy river to be a threatening and voracious goddess, which spells out a deep rift between pilgrims and villagers when it comes to the nature of divinity. I will engage with this rift in detail in chapter 7, where I unpack ritual and gendered dimensions of environmental relations sustained in visions of risk and protection.

Colony Para, the resettlement colony Ganesh calls home and where I conducted the bulk of my fieldwork, sits in the shadow of the pilgrimage complex. Home to some 180 households of varying sizes, it is one of five resettlement colonies dotted along Sagar Island—the other four being Harinbari, Bankimnagar, Jibantala, and Beguakhali (see map 1.1). All these colonies were established more than thirty years ago and turned into crowded settlements since. And all of them mirror the demographics shaping the island and the Indian parts of the delta front as a whole: a Hindu majority, composed of lower castes of the OBC category—that is, Other Backward Castes—to middle castes, locally referred to as "general caste," and complemented by a Muslim minority, made up almost entirely of low-ranking Sheikhs. Taking the number of households as an indicator, my census shows Colony Para's Muslims to amount to roughly 30 percent and Hindus to 70 percent, again mirroring district-level demographics. Occupations and livelihood appear of similar types, and diversities, as is being reported from all parts of the Sundarbans. Most people engage in seasonal mixes of providing services in the pilgrimage complex, construction activities, or fishing, backed up by agricultural work on their own fields. The considerable investments needed prevent all my interlocutors from owning their own fishing trawlers. Instead, many are employed on trawlers owned by investors or merchants, and work either as day laborers or for the duration of a season, or on similar terms on dry fish operations, drying low-quality fish for the domestic market in camps set up by investors. Members of the poorest households would also catch tiger prawn seeds (*meen*) or small fish with retrofitted mosquito nettings along the shores and banks, selling the former to middlemen fueling the highly disruptive global prawn industry (Jalais 2010a; Paprocki 2019).

Circular labor migration (Massey 2009) appears an essential strategy to make ends meet in virtually all households across all my field sites. It might be in the form of regular circular migration, such as women staying as servants in urban middle-class homes (Ray and Qayum 2009) or men spending the better part of every year on poultry farms in places as far as Kerala on India's southwestern tip. Most often, however, it takes the form of groups of men, sometimes accompanied by women, taken by middlemen to construction sites across Eastern India on shorter assignments, ranging from a few weeks to a few months.

Given the shifts in governance policies and the mounting scarcity of land, the process of official land distribution has long been closed, and displaced islanders now face bleak outlooks. Following relations and stories along this morphing coast allows me to put the experience and afterlife of coastal erosion, as well as the politics and poetics of resettlement, to ethnographic scrutiny. This book does so by drawing on seventeen months of fieldwork, spread out across ten years, using the full spectrum of ethnographic research, including participant observation, various forms of interviews, and focus groups, enriched by the analysis of textual and graphic material all along the way.

The disruption and liquification of homes and villages on Sagar Islands' worst-hit stretches may—as this book argues—be particular, but are not unique. Across the Sundarbans, coastal erosion is endemic. Coastlines morph rapidly, given the triple pressures of waves, currents, and tides in a rising sea. The demise of rivers due to the combination of geomorphological shifts and anthropogenic interventions at various scales only aggravates the situation, as I show in chapters 3 and 6. For decades now, more land has been washed away than what is deposited. Where substantial land was accreted, however, as on the now rather massive island of Nayachar, state authorities precluded the settlement of farmers for the benefit of, first, industrial complexes and, since this provoked resistance, salt-water aquacultures.

Against this backdrop of erosion and inaccessibility of other lands, long coastlines have become zones of despair, impoverishment, and failed hope. It is a development that leaves an ever-increasing number of villagers stripped of their lands and displaced from their homes. Being endemic, coastal erosion is particularly swift and visible in a number of hotspots. Botkhali is among them, as are the islands of Moushuni and Ghoramara nearby (Aditya Ghosh 2017; Harms 2015; Danda 2007; Centre for Science and Environment 2012). Furthermore, coastal erosion never stands still, as I show in this book. That is, as water encroaches, shorelines crumble and people retreat, zones of erosion wander and engulf landscapes and lives hitherto removed from the shore. There is little doubt that what now occurs in Botkhali will eventually descend on inland villages, the next targets once Botkhali is gone. In effect, the sea is literally encroaching upon people, one layer at a time. And many of my interlocutors look back on pasts shaped by micromigrations, receding with the shore, rebuilding and again losing their homes—frequently more than five times.

But if erosion is endemic in delta fronts and if the amphibious character of the land is a fact taken for granted, where does trouble emerge, and when does disaster strike? It lies, I suggest, in a double bind concerning the nature of land. On the one hand, settlement patterns staunchly defy the amphibious, as I detail in chapter 4. On the other, land has ceased to be available. In the course of settlement, most jungles have been axed and swamps drained. Others have been fenced off as protected areas. Still others appear earmarked for industrial development. Now, the time-honored strategies of moving on and finding another swamp to drain and colonize lead nowhere. Land is exhausted and forests are beyond reach, the agrarian frontier melting away.

To suffer coastal erosion is not unique to the larger region. Similar can be said about the majority of all coastlines in the vast Bengal delta stretching from here eastward for some one thousand miles and covering most of the coastal tracts of India's West Bengal and of neighboring Bangladesh. These are extremely dynamic shores, subject to complex, local geomorphological scenarios (Brammer 2014). And while accretion seems to increasingly outweigh erosion the farther one moves to the east, coastal erosion affects populations all along the shoreline, at least in pockets. Zooming out even further, coastal erosion is now causing suffering across the low-lying, densely populated, and rapidly urbanizing coastlines of Asia and the world. Scholars, activists, and journalists highlight the plight of living in, and living with, disappearing shorelines in circumpolar regions and Small Island Development States (Marino 2015; Maldonado 2018; Rudiak-Gould 2013). With the Bengal delta expected to be hit particularly hard, it makes sense to conjure up a map of zones of coastal erosion and land loss dispersed on the globe. The morphing of the coast also intersects with state neglect, endemic poverty, and extremely high rates of population density. Exploring how people live with eroding—and liquifying or shrinking—landscapes in the Sundarbans, this book provides a nuanced account of localized "environmental suffering" (Auyero and Swistun 2009) that has import for coming to terms with planetary injury.

Multiple Sundarbans

The islands are the trailing threads of India's fabric, the ragged fringe of her sari, the *āchol* that follows her, half-wetted by the sea. They number in the thousands, these islands; some are immense and some no larger than sandbars; some have lasted through recorded history while others were washed into being just a year or two ago. These islands are the rivers' restitution, the offerings through which they return to the earth what they have taken from it, but in such a form as to assert their permanent dominion over their gift. The rivers' channels are spread across the land like a fine mesh net, creating a terrain where the boundaries between land and water are always mutating, always unpredictable. . . . There are no borders here to divide fresh water from salt, the river from the sea. . . . The currents here are so powerful as to reshape the islands almost daily—some days the water tears away entire promontories and peninsulas; at other times it throws up new shelves and sandbanks were there were none before.

—Amitav Ghosh, *The Hungry Tide*

Considering life amid environmental injury on extremely low-lying shorelines, I traverse the immediate coastal fringes, which, despite being situated just south of the East Indian metropolis of Calcutta/Kolkata,[3] seem worlds apart from the city. This is an area of superlatives and extremes. It forms a part of the world's largest delta, housing the largest-remaining uninterrupted stretch of mangrove swamp on earth, home to the highest

---

3. In 2001, Calcutta was renamed into Kolkata. I use the name as applied at the respective time referred to.

number of South Asia's much adored royal Bengal tigers. To make room for feline beasts and mangroves, both regarded as obnoxious and awe-inspiring at the same time, colonial governments created schemes of protecting their habitat (Greenough 1998). Continued by the postcolonial states of India, East Pakistan, and, later, Bangladesh, these schemes resulted in the establishment of substantial protected areas at the heart of the delta's coastal tracts (Greenough 2003; Jalais 2008, 2010b). True to the notion of "fortress conservation" (Brockington 2002), environmental protection here continues being a militarized affair, with fences, patrols, fines, and incarceration. Everyday violence is carried out against unregulated forest users, and came to the boil most destructively with the massacre of Marichjhapi in 1979 (Jalais 2005; Mallick 1999).

This is an area where tigers are gods and gods are tigers, embedded within Hindu and Muslim lore. It is an area where land cannot be trusted. Colonial authors were irritated by these shifting landscapes, halfway between liquid and solid (see chapter 4). Novelists turn to them for their marked fluidity, defying easy categorization and, as such, housing strange creatures. It is no coincidence that the protagonist of Salman Rushdie's delightful novel *Midnight Children* nearly goes mad in these swampy lands, or that Amitav Ghosh (2004, 2019) keeps returning to the Sundarbans. The landscape is brimming with monsters and stories, enchanting not in spite of but precisely because of its shapeshifting nature.

Trees growing in salty water, archipelagoes rising and sinking with the diurnal tides, and expanses of monotonous salty green rising from amorphous muddy lands embody the tattering ends of some of Asia's mightiest and most revered rivers. Long rendered obnoxious and vilified (Sarkar 2010), the protected heart of the Sundarbans now attracts an ever-growing number of visitors keen to admire the greenery, big cats, and what appears to be the lure of the jungle right at the gates of the Bengali metropolis.

In descriptions of the Sundarbans, references to any superlatives are interlaced with references to marginality. Its soggy, remote terrain also figures as a poorhouse, full of swamps and uncertainty, coupled with a sense of chronic existential dread. Entering the chronicles of modern history as a destination for the destitute masses squeezed out of rapidly populating land-stressed Bengal, as a colonial dream where immense profits can be plundered from footloose settlers (Sarkar 2010; I. Iqbal 2010; S. Bose 1993, 2007), it has been reframed as a zone left behind from national aspirations by virtue of its inaccessibility, the regular devastation wrought by cyclones and surges, or the stupefying poverty of its soils. For decades, geomorphologists have warned that the area is unsuitable for human habitation, endorsing bourgeois contempt at misers competing with tigers for scarce resources within and along protected areas (Fergusson 1863; Jalais 2008; Bhattacharyya and Mehtta 2020).

But if conservationist or tourist gazes are firmly directed toward the heart of the jungle in the sanctuary, this overshadows life beyond the park and the jungle within the same ecotone. Contemporary usage patterns of the label "Sundarbans" are hence somewhat treacherous. Indeed, my conversations with urban Bengalis would frequently affirm the bias that underpins much scholarly work on the active delta. Upon explaining my research, I frequently encountered resistance, culminating at times in the suggestion to refocus my study away from the immediate coastal fringes of the delta and toward the *real* Sundarbans—that is, the mangrove jungle shared by big cats, impoverished farmers, and fishermen.

In other words, this book starts from the premise that there are further Sundarbans situated around the core jungle, its buffer zones, and the habituated edges layered around all of them. This premise is well founded, as much larger areas than the protected zone and its immediate environs have historically been considered parts of the Sundarbans (Ascoli 1921; De 1983; Pargiter 1934). They continue being classified as such by many of their dwellers and by bureaucratic routines. More than that, the formerly swampy and now increasingly fragile, ragged islands and tongues of land embodying the archipelago south of the metropolis, between city and sea, share several trajectories. They are of a similar material makeup, characterized by the young accumulation of footloose sediments, stable enough to be called land, yet volatile enough to be quickly washed away; fresh enough to make them eligible for agricultural lifestyles, yet never far enough from the reaches of marine waters threatening agricultural prospects, much to the despair of people who consider themselves farmers. If living surrounded by water defines much of Bengal's floodplains, with all the awe, sublimity, and terror in the eye of past or contemporary European observers (see, e.g., Gardner 1991), the Sundarbans share in the ubiquity of *brackish* water. Furthermore, the archipelago that makes up the Sundarbans is marked by being exceptionally hazardous. A glance at the literature reveals that hazards have been and continue to be diverse, and not all of them affect all parts in the same way (see, e.g., Bandyopadhyay 1997; H. S. Sen 2019). My research partners were relieved to say that the times when they had to fight off tigers or crocodiles are long gone, that these deadly creatures would continue troubling people closer to what remains of the forest.

Still, the days of tigers are well remembered, alive in folklore, and written into place-names all around. Likewise, a few hazards have diminished in importance, such as earthquakes (Nath, Roy, and Thingbaijam 2008) or fevers (Greenough 2003; Nicholas 2003; U. P. Mukherjee 2013), which gripped colonial observers. Most importantly for this book, however, the entire archipelago continues to be vulnerable to a range of water-related hazards. Journalists frequently portray the Indian parts of the Sundarbans as disaster prone, mirroring diagnoses of Bangladesh as a whole (Haque 1997). Of late, the islands are referred to as a climate frontier, as a poster child of climate change, or as home to India's or the world's first climate refugees (see, e.g., Cockburn 2010; AFP 2019; Chowdhuri 2018; J. Gupta 2018). In line with that, West Bengal's coastlines, of which the overarching majority fall into the Sundarbans, are the shores most rapidly affected by coastal erosion across all India (Kankara, Murthy, and Rajee-van 2018). Finally, sea levels appear to rise more rapidly here, with the mean sea level rising in Sagar amounting to 5.74 mm per year between 1948 and 2004, distinctly above the global average of 3.3 mm per year for the same period (Unnikrishnan and Shankar 2007, 306). As a whole, it appears as what Jason Cons (2018) calls a "climate heterodystopia"—a place of difference, yet of a dark and troubling kind, a place of experimentation and witnessing from afar what the future has in store.

Even if these other Sundarbans have made headlines recently, as actual or potential victims of sea level rise and a range of further anthropogenic environmental degradations, the available literature is still strongly oriented toward the habituated edges circling around the tiger habitat—to the exclusion of millions living on unstable islands in the active delta or the immediate delta front. This book contributes to correcting the picture, demonstrating that the Sundarbans are a diverse and stratified terrain. In her study of tiger relations close to the sanctuary, Annu Jalais (2010b) speaks of hierarchical orderings of islands—the closer to the city, the higher, the farther off, and the lower—mingling urbanization, development, and value in uneasy ways. On top of that, many islands in this vast archipelago are marked, as I will show, by divides between edge and interior. The spaces of slightly wealthier people who own slightly more lands stand out next to a terrain of lesser prospects and higher vulnerability. Such differences can be mapped alongside the dividing lines of political persuasion in a terrain infamous for its deeply entrenched party politics (Ruud 2000; Mallick 1993). Divisions of caste and creed further complicate the picture, yet rarely provide the sort of stark divisions we know from many other South Asian societies. Overall, the volatile frontiers of the delta have been known as regions where caste and creed lose importance against what has been described as a sense of egalitarianism in the face of hazardous natures and troublesome pasts in which paupers were swept into fever-ridden jungles (I. Iqbal 2010; Nicholas 2003). Thus, those who call the swampy frontier home tend to be poor lower-caste Hindus and poor low-status Muslims (for more details, see chapter 4), generally enjoying the absence of strained relations or communalist sentiments. My account stops short of providing yet another portrayal of contemporary Bengali society and politics as if it were untainted by the logic and practice of caste. Caste certainly underpins party politics and the social life of development policies in practice (Chandra and Nielsen 2012; Chandra, Heierstad, and Nielsen 2015). Caste also helps to account for some of the injustices in sustaining environmental injuries (M. Sharma 2017). Yet, as I will argue, the frontier region of the Sundarbans offers promising ground for rethinking the relation of caste and class in a crumbling frontier society.

I will explore these dimensions at length. Suffice to say for now that against these cascading differences, one might think of the archipelago as one of multiple Sundarbans. There are more than one Sundarbans and "less than many," to adapt Annemarie Mol's (2002, 52) phrasing, and they all hang together in tension. Coastal erosion is messing up such neat socio-spatial arrangements. In their brute and tumultuous ways, encroaching and eroding shorelines move the coastal fringe toward the islands' interiors. Neighborhoods are impoverished and rendered frontier zones. Meanwhile, in many cases—what by local standards count as—wealthy peasants become landless have-nots over the course of years and decades.

Anticipating Loss and Distributed Disasters

We are living through an extended phase of planetary injury and fragmentation. More than ever, survival seems at stake. Mourning the disappearance of life-forms and whole ecosystems—not long ago a pastime limited to activists and scholars—now permeates public debate and intimate encounters with nature. Scholars at the intersections of the social sciences, psychology, medicine, and philosophy call attention to specific modes of suffering that require being addressed through novel categories. Albrecht (2019) speaks of solastalgia, Sarah Jaquette Ray (2020) of climate anxiety, while Bruno Latour (2018) identifies a sense of displacement without moving as a marker of the present.

This predicament manifests, I contend, in intricately knotted temporalities. As always, it is one of path dependencies, historical burdens, and uneven responsibilities stemming from past actions. For one, the Anthropocene's anthropos is a splintered and torn, and more or less privileged and more or less responsible, conglomerate—not a unified actor—that must be scrutinized with the help of the analytical toolbox of postcolonial studies and political ecology (Baldwin and Erickson 2020; Gibson and Venkateswar 2015; Moore 2016). A lot has been written on excess and accumulation in the worlds of environmental decay. In this age of fallout, past emissions accumulate in the air and in rivers, as bodies accumulate toxins (Masco 2015; Murphy 2015). Industries and cities clog landscapes, their excesses embedded well into the future (Stoler 2013; Gordillo 2014). In addition to the excesses of emission ousting and poisoning, the globe also sustains a biodiversity implosion and the disappearance of ecosystems or landscapes (see, e.g., Kolbert 2014; Hallmann et al. 2017)—in short, the thinning of life. Irrespective of considerable successes in the field of environmental conservation, environmental degradation intensifies steadily. We now live through collapsing ecosystems and vanishing worlds. Future possibilities diminish, and the present turns stale against the background of what we know it used to be. This predicament of thinning futures accentuates temporalized, increasingly fragile encounters with nature marred by an anticipation of want. Alongside apparitions of lost species and the monsters of runaway growth (Tsing et al. 2017; Tsing 2015), people increasingly look at valued parts of their material environments—say, individual plants, species, landscapes, or meteorological appearances—as if they are disappearing, even as if they are already gone. This calls forth a sense of bidding farewell, often accompanied by an ethics of enjoying that which is disappearing as long as it lasts. In such takes, the contemporary appears as the ghost of an even more precarious future.

Meanwhile, when approached differently, our present is saturated with nostalgia, particularly when social movement and writers imply that in some ways it is already too late, that climate change and mass species extinction have already arrived (Franzen 2019; Farrell et al. 2019). This book contributes to coming to terms with this moment of anticipating futures marked by loss, of relating to the present as a stale reminder and a ghost to become. When I think landscapes or people through the lens of future ghosts, I do not mean to subscribe to visions of the wholesale inundation of entire coasts or deltas amid sea level rise, nor do I intend to subscribe to discourses of victimization and the erasure of agency. Such visions would disregard coastal heterogeneity, resilience, and adaptive interventions that hold the potential to transform battered coastlines into mosaics of endurance. While this is generally true, a long history of neglect and the sidelining of comprehensive adaptation measures along Bengal's rural coasts make the thinning of futures and a turn to the ghostly appear very likely.

In turning to the disappearance of land, homes, and swamps on eroding coasts, this book accounts for one of the most extreme articulations of environmental degradation. As coastal erosions and ensuing population displacements are now a widespread reality on low-lying coasts across the world, this is an essential endeavor. It is also imperative to understand and act upon a range of environmental transformations that appear to be unrelated yet share similarities with respect to textures of morphing and modes of navigating. This materializes in processes of desertification and the poisoning of landscapes, as well as continued industrial and infrastructural developments in the age of mushrooming population density. Ethnographically scrutinizing the ways in which the rural poor navigate eroding terrain amid social constrains and fragile hope, the following chapters offer an account of the multiplicity of disappearance, accounting for temporal rhythms, multiple natures, technological interventions, and political dynamics.

In attending to long-deforested and drained landscapes at the outer marine limits of the delta, and being attuned to the shifting, messy nature of life enmeshed in erosion, this book makes the case for using loss as a prism to think coastal futures. This is not to say that coastal dwellers are easily overpowered and victimized, or that loss afflicts all humans proportionately. Instead, this book posits that we must remain alert to the ways in which loss as an anticipation and as a condition informs the way the present is dealt with and the way futures are made on sinking and battered coasts. Meditating on loss, Judith Butler notes: "Loss becomes condition and necessity for a certain sense of community, where community does not overcome the loss, where community *cannot* overcome the loss without losing the very sense of itself as community" (2003, 468, emphasis in original). To think coastal erosion through the prism of the community of loss captures not only the totality of devastation, but first and foremost this shared sense of bereavement. At the same time, it invites us to attend to the state of absent materialities and the conflation of diverse temporal modes through the specter of loss. Focusing on loss and suffering as a shared predicament helps us to grasp, I contend, instances of solidarity, care, and tolerance essential for navigating the onslaught by the sea. This is not to suggest that instances of generosity and altruism can surmount the toll of past tragedies, but rather that it helps to account for the moments of care that crop up amid competition for scarce resources (Werbner 1980, 1996; Khan 2022).

Coastal erosion overflows the conceptual apparatus applied in research and practice on disasters. Navigating terrain as it gets undermined or washed away not only entails a process of relearning and readjustment. It also requires a kind of accounting for and of tackling these devastating developments. Turning our attention to lives lived amid encroaching seas, liquefying lands, and collapsing hopes, this book makes the case for the analytical and practical value of attending to the middle ground between endemic crisis and specific catastrophe, between chronicity and disruption. Coastal erosion appears, as I demonstrate in chapter 3, as a distributed disaster spread out across time and space. It is one of the concerns of this book to help to reorient the way disasters are thought about, discovered, and combated on a planet in peril.

To address coastal erosion means to get attuned to drawn-out processes. Plots get washed away across a span of months or years. Decisions to abandon homes are frequently not taken in a rush—they are years in the making. Villages and islands disappear within decades, as everything that cannot be unanchored slowly and relentlessly gets consumed by water. Yet, ruptures animate this violence of attritional changes, coalescing into moments of heightened dangers and intensified suffering. Cyclones not only are a recognized threat to survival, they also magnify—as I show in chapter 3—the reality of rather ordinary coastal erosion. Surges overrun shore fortifications, flooding, morphing, and washing out villages. Tidal waters enter freely through breaches, and homes are washed out as terrain becomes marine. Such an entanglement of event and process is reflected on the level of individual navigation of eroding terrain, spelling out times of being, as I call, in the fold of erosion interspersed with moments of flight, quick dispersal, and letting go. If we zoom out, what appears as a highly individualized process of being ousted from one's home by a stalking sea turns out to be shared as an experience by swaths of the population. As vast shorelines are affected by dramatic morphing, waters close in on people spread across the archipelago, affecting some coasts and sparing others. In time, people move out of flood zones often only to be grabbed again. My notion of the distributed disaster thus occupies the middle ground without losing sight of either pole. I claim that to attend to this middle ground is critical for a better understanding of disasters, and thinning, in this moment of planetary peril.

Mud, Water, and People

In coastal erosion along the Bengal delta front, event and condition intertwine. Seen from afar, the morphing of coasts has long been considered the norm in the Sundarbans. For current scientific publications, this is a textbook case of delta dynamics. Accretion and erosion go hand in hand in crafting an amphibious terrain that always seems likely to expand, shifting the delta's seafront elsewhere.

Not erosion per se but its increased pace and scale render it a deeply problematic affair—indeed, an abnormal and malevolent process. Fear of submergence runs like a thread through modern thought and beyond, as the preoccupation with sunken cities or continents, such as Atlantis or Lemuria, demonstrates (Ramaswamy 2005). It is the sheer speed and the preponderance of erosion in the westernmost parts of the Bengal delta that make them, I suggest, a laboratory for living with sea level rise and a smokescreen for anxieties about supposedly shared futures from afar. Concomitantly, it makes headlines as a disruption—disrupting rural riches, and lives, at the fringes.

Geographers argue persuasively that the actual shape the delta takes is deeply anthropogenic. The long history of deforestation and intense agriculture in the extensive drainage areas, have, via soil erosion, increased the amount of sediments swept downriver for centuries (Chakrabarti 2001), thus facilitating the blooming of landmasses at the rivers' mouths. Likewise, the colonial obsession with checkering floods—now identified as a threat rather than a replenishment—has greatly contributed to the siltation of rivers, as sediments were blocked from being spilled across emerging landmasses with floods and tides. This siltation again has choked whole channels, blocking waters and forcing new pathways, which made the emblematic madness of the rivers human administered to a significant degree (Dewan 2021; D'Souza 2006).

The Ganges delta in Bangladesh and adjacent East India have been critical sites for researching erosion at the land-water interface. Considering the erratic nature of South Asia's rivers, which ferociously swell during monsoons and transform the plains with a perplexing arbitrariness, it is little surprise that riverbank erosion in the delta was a major cause of concern for colonial authorities (D'Souza 2006). However, the literature appears to be strongly tilted toward inland estuarine lifeworlds in flux, at the expense of what I understand to be eroding landscapes on threatened seafronts. In Bangladesh, generations of geographers and their allies investigated the social consequences of land erosion by major rivers (Lein 2009; Zaman 1989, 1991; Schmuck-Widmann 2001;

Hutton and Haque 2004, 2003). Dispute ensued about whether this was something calamitous or simply business as usual in a young waterscape where people are well accustomed to living with volatile and rarely controllable rivers, and whether land loss sustained by rivers' shifting courses led to displacement and strife or was cushioned by rules attributing newly emerging lands to displaced people (Lahiri-Dutt and Samanta 2007). This debate is mirrored on the other side of the border dividing Bengal and India since the partition. Here, one series of publications explores riverbank erosion explicitly as a disaster, a disaster that liquifies assets and pushes farmers into destitution and displacement (Rahim, Mukhopadhyay, and Sarkar 2008; S. Iqbal 2010). Another set of studies emphasize the fluidity of riverine worlds, addressing the practices of adaptation and flexibility required to live on ephemeral, constantly shifting lands. Intriguingly, this latter body of scholarship mostly turns away from settlements by the side of the river and refocuses instead on life between the banks, focusing on highly instable sandbars and islets that harbor specific forms of life, *chars*. People appear to move along with fickle floating lands, showcasing mobilities, economic rhythms, and forms of social organization suitable for engaging in, as Kuntala Lahiri-Dutt and Gopa Samanta (2013) evocatively put it, dancing with the river.

Thus conceived, *char* life calls attention to mobility as adaptation, to the frontiers of terrestrial life-forms, and to the limited practicality of organizing property in terms of the rigid, supposedly stable grids and land registers seen in so-called modern states (D'Souza 2006). As an empirical entity and an inspiration to make up more environmentally modest futures, the *char* keeps on traveling. It is being discovered far beyond the Bengal delta (N. Sinha 2014; Albinia 2009; P. Sen 2008). Yet its presence, and its salience in thinking about amphibian life, is restricted, I argue, by specific environmental relations as well as by approaches to land and self. *Char* life, in other words, is decidedly riverine, not estuarine or coastal. To dance with the river requires water to be fresh, or *mishti*, sweet, as Bengalis would have it. To people of this delta, freshwater floods are not per se calamitous, they can also be invigoratingly fertile, creating new lands that are almost immediately fit for cultivation (Lahiri-Dutt and Samanta 2007). It's different at the seafront—here floods by uniformly brackish waters ravage harvests, with newly emerging land needing time and sustained effort until its salinity can give way to sweetness, the precondition for agricultural use. To live in such a state of flux and to move along with floating lands is unappealing and even morally suspicious to large swaths of society (Schmuck-Widmann 2001). Literature on *chars*

alerts us to all too familiar distinctions articulated by bank people or village people, who chastise floating people for living morally hazardous
lives, as well as by *char* settlers who pride themselves on living lives less
touched by restrictive social norms and etiquette.

In being concerned with islands that are also in flux, this book still
discovers lifeworlds that contrast decidedly with *chars*. Given the salty
conditions, *char* life is just as unfeasible on the seafront as it is on the
estuarine reaches. Islanders subscribe, as I will show, to a view of land that
revolves around a modest level of stability and reliability. In this, colonial
regimes of settlement coalesce with Bengali notions of rural beauty and
the good life. Taken together, however, they underline the point that erosion on the seafront has a different texture than it has in the freshwater
worlds far up the river, as I demonstrate in detail in chapters 3 and 6.
Complementing the literature on a somewhat playful morphing of amphibious islands deep in the interstices of monsoon rivers, I reveal life on submerging coastal landmasses as it is lived by people who must adhere to an
ethics and aesthetics of terrestrial life. It is a life permeated by loss.

Landscapes implicated in coastal erosion call into question how we
think mobility. Nature and material environments themselves seem to be
in a whirlwind of movement, their impermanence and transformation
dramatically accelerated as shifting shorelines swamp and remold terrain.
In a sense, the world is out of joints and terrestrial thinking is numbed
by the force of waters operating in the mode of excess, spilling, soaking,
and submergence. In contrast to parched regions where water's incremental absence or dispersion is seen to thin out places and leave people
stranded, I engage with lifeworlds as they turn untenable through the
overwhelming incursions of water of a disruptive, saline kind. In both
cases, matter's mobilities, fluidities, and cascading transformations, accelerated in configurations of crisis, unveil specific patterns of human
mobility. It will not suffice to situate such mobilities only against a backdrop of environmental transformation—for the routes and trajectories
of human mobilities are clearly demarcated by layers of political and economic constraints. Nor will it suffice, as I argue in this book, to conceptualize mobility in this shapeshifting world within a dyad of staying and
fleeing. They come entangled, and one feeds into the other.

The debate concerning the overlaps of environment and migration
appears to be in gridlock. Discussions continue unabated on numbers,
on their reliability and suitability to forecasts (see, e.g., Kelman 2019).
Very rough estimates travel back and forth between scientific publications, policy papers, and journalist articles, sometimes sticking even after

being discredited as bloated or unsupported by evidence (Hunter, Luna, and Norton 2015; Lübken 2012; Oliver-Smith 2012; McLeman and Gemenne 2018). This is not to say that the issue of forced migration in the context of environmental changes or disasters is not of numbing magnitude. It certainly is. And the future appears bleak. On a different note, we need to concern ourselves with what such numbers can do. If all goes well, quantifications seduce by providing a sense of hard data, of truthfulness and the clout of science (Merry 2016). But they are easily harnessed by xenophobic visions and conceal as much as they reveal, both on the criteria they rely upon and on the problems they claim to represent. Finally, numbers and curves are naturally devoid of the empathy and solidarity that comes with adding faces and concrete experiences. This book bets on ethnography. It attempts to steer clear of debates on numbers for the most part and instead highlights the way people move along with shifting coastlines amid social constraints. In doing so, the following pages answer the urgent need for detailed accounts of real-world experiences of environmental displacement on a warming globe.

Furthermore, the debate on environment and migration is imbued with deep rifts on what counts as its object. Quarrels abound on two fronts. On the one hand, scholars provide competing categorizations intended to capture and systematize divergent migration trajectories. Labels cohere with timings (before, during, or after a shock), the degree of permanence of outmigration (when they distinguish between flights or relocation), or the amount of deliberation involved in decisions to move (Suhrke 1994; Black 2001; Castles 2002; Black, Adger, et al. 2011; Morrissey 2012). Much of the debate takes sedentariness as the implicit norm: being anchored in place appears as ideal and mobility as problematic. This resonates with inscribing populations in place at the expense of embracing the fact that migration and mobility have also always been common practice (Appadurai 1988). Subsequent turns to thinking migration in itself as a form of adaptation to environmental transformation have ameliorated this problem to a degree (Black, Bennett, et al. 2011). But this book reveals that the borders between mobility and staying put are not merely porous but collapse in environments characterized by overwhelmingly mobile matter. This distinction fails lifeworlds where engaging in micromigrations following seawater incursions is, paradoxically, a tactic of staying.

My account reveals complex and dynamic patterns of mobility enfolded reiteratively in the fluctuations of this muddy waterscape. These include, but reach beyond, the entanglements of mobility and immobility we know from the worlds of marginalized labor migration. As it turns out, many people are mobile precisely in order to stay put. I elaborate in chapters 4 and 5 on ensuing circuits and risks, relying on the conceptual apparatus of translocality (Greiner and Sakdapolrak 2013) and migration as adaption (Black, Bennett, et al. 2011). In addition, I discover another mode of being mobile in order to stay put: going through environmental displacement, people are forced to move, yet in large numbers stay in the vicinity. Retreating from the shoreline, for a long time many move not completely away or out of harm's way. This contrasts with, I argue, much of the migration literature, which evokes swelling ranks of footloose populations crossing continents or even nation-states. What I dub "micromigration" occurs well below the radar of migration regimes and beyond resettlement or planned relocations schemes. Staying put also cannot be explained simply as an outcome of being too poor to move. Accordingly, I analyze short-range migration as a strategy driven by the attempt to tap into desperately hoped for rehabilitation schemes. It is a strategy buttressed by a sense of belonging and the tacit permission of former neighbors, which unfolds as a process of suffering when the shoreline draws near all over again.

Finally, scholars struggle to specify the role and relevance of environmental processes and events within migration trajectories (El-Hinnawi 1985; Klepp 2018). To pinpoint actual triggers and root causes is, of course, critical for any deliberation of legal rights and obligations. However, it fails to encapsulate the complexity of actual decision-making among mobile actors or the distributed and complex—oftentimes invisible and incremental—impacts of environmental degradations. It often remains unclear what role environmental degradations or disasters play within decisions to move—specifically among marginalized populations navigating multiple interlocked crises across the spheres of health, income, land tenure, and education (see, e.g., Klepp 2018; Oliver-Smith 2012; Piguet and Laczko 2014). On the impoverished, hazardous, neglected, and densely populated Bengal delta front, such complexities ensue. A growing number of people move across and out of the delta for various interlocking reasons (Aditya Ghosh 2017), frequently in the aftermath of the latest storm. But here again the extreme event turns out to be, as I will argue, the mere tipping point in altogether unstable lifeworlds and somewhat overdue decisions. I explore this figuration in detail across chapters 3 and 6.

Ironically, the texture of coastal erosion along a choking estuary in a crowded delta relieves me of much of the conceptual troubles. As shorelines encroach on homesteads, embankments collapse, and all that is solid is washed away, there is ultimately little room for debating whether nature figures large in mobility trajectories. When the floor underneath one's feet disappears, taking with it the home and the most fundamental constituents of livelihood, displacement ensues in a straightforward manner. Across the chapters of this book, I take care to situate erosion within the contexts of history, political ecology, and practice. However, this does not denigrate the immediate rawness of the experience through which my friends and interlocutors saw themselves ousted by turbulent waters. In this sense, coastal erosion of this magnitude may be one of the very few dynamics allowing one to use the label "environmental displacement" in a nonqualified way.

Furthermore, I like to think of the ousting by brackish waters as a dis-placement, attempting to capture therewith the immediacy and force of liquifying places as well as the horror involved in seeing the ground beneath one's feet dissolving. Clumsy as it is, I occasionally use the hyphenated term across my chapters. "Dis-placement" calls attention to the ambiguities and difficulties sustained by shifting shorelines and the swamping of earthbound human life as well as to the shattering of terrestrial thinking by an encroaching sea.

To speak of dis-placement means also to zoom in on experiences of forced mobility that diverge from the much more familiar contexts of war, strife, and ethnic cleansing (see, e.g., Butalia 2000; Kaul 2001) or urban beautification (see, e.g., Tarlo 2003; Ghertner 2011). The story I am telling in this book only rarely touches upon fences, camps, bulldozers, or weapons. Rather, I follow the gradual dissolution of space and the fading away of places. Through the prism of dis-placement, I ask for the social life and afterlife of lost homes, lands, and landscapes. How, in other words, are retreating shorelines and vanishing islands navigated by marginalized communities? How do islanders live with and make sense of the slow violence inflicted by a sacred river? What kind of disaster is this, and what is its texture? How do people cope with drawn-out changes and deteriorations amid state neglect, and what kind of futures do they anticipate? How are sunken places remembered? How is it to live on land that is thinning, already bearing the mark of the sunken and gone? How is it, in other words, to live on an already foregone land?

Tracing Disappearance along Knotted Temporalities

On that hot day in June 2019 when Ganesh and I were inspecting changes and relearning the landscape, we did so by contrasting older forms of the coast with the present. The transformations we zoomed in on would certainly be lost to most maps. Not because they were too miniscule or irrelevant to mapmakers, but rather because the present scenario had only a short lifespan. It was in the process of being eroded as we walked it. And as this book goes to print, it will most likely have been washed away for good, the village of Botkhali wiped from the face of the earth.

What remains of Botkhali, or indeed of other "dead places" (Read 1996, ix) I am concerned with in this book, is hardly visible to the eye. To experienced navigators, certain currents that reveal bumps in the sea-floor will give away the location of former islets. Displaced islanders will point into the watery expanse, expounding on the precise whereabouts of the house they grew up in or the trees whose shade they rested in before both disappeared. Similarly, sturdy infrastructural objects, such as sluice gates or trunks, might continue to signal former contours of the island by sticking out of the water before they also fade from sight. On most accounts, merely a watery expanse remains. As the estuary constantly gnaws away at the land, pushing through embankments and into homes, a rift deepens between the stable coastlines found in maps and the fluid, retreating, morphing character of coastlines in terrains such as the Sundarbans, filled only by water and memory.

In light of this rift, this book turns to memories of the past and their narrations as critical resources. Narrations of the past have been the staple of historians and anthropologists, in oral history and beyond. The sphere of intimate recollections, once understood as the private foundation of selfhood, has become susceptible to all kinds of interventions (Halbwachs 1992; Berliner 2005; Augé 2004; Connerton 1989). A number of studies have interrogated the ways in which the past is deemed meaningful or used as a resource to bolster critical accounts of the present (Gold and Gujar 2002). After all, as Michael Lambek (1996, 240) notes, memory is "never out of time and never morally or pragmatically neutral." Students of disasters have likewise shown that remembrance is a key resource for coping with disruptive events and a vehicle for negotiating meaning and attributing blame (Gray and Oliver 2004; Oliver-Smith 1999a; Simpson 2011). In all these accounts, memory appears as malleable.

To trace disappearance, then, is an exercise in working with, and working against, forgetting—that ubiquitous and by definition invisible yet socially informed counterpart to functional memory (Connerton 2008, 2009; Esposito 2008). Marc Augé (2004) binds erosion and forgetting together in a marine metaphor that shrewdly grasps the workings of memory. He writes, "memories are crafted by oblivion as the outlines of the shore are created by the sea" (2004, 20). I nurture this concept further by turning to memory and forgetfulness as I excavate environmental relations in this frontier zone. Memory opens up a window to understand how islanders frame a world dramatically in flux, how they navigate displacements and the everyday challenges of living surrounded by a hostile waterscape, and how disappeared lands may persist.

Anthropology has long been concerned with disappearance and loss, cataloging bodies of knowledge and technologies as they disappear in the onslaught of modernization. Lévi-Strauss's (2012) haunting *Tristes Tropiques* is emblematic in combining such an approach with cultural critique. While these studies were literally working against the loss of Culture in its capitalized form, present studies rather attempt to make sense of the ways people live through decaying material environments (Stoler 2013; Gordillo 2014) or how absent things, places, or people are rendered present in practice (Bille, Hastrup, and Sorensen 2010; Meyer 2012).

Exploring shoreline morphing and here, specifically, coastal erosion entails engaging with knotted temporalities of disappearance. These come in three modes. The first—I have already introduced it as *relearning*—involves localized future making in light of assessments of coastal erosion. The second revolves around the complex temporal texture of erosion, involving what I introduce as *distributed disasters* spread unevenly in time and involving certainty about onslaughts and losses. The submergence of islands in the Sundarbans also renders them, to onlookers from afar, sites to witness, investigate, and depict coastal futures as present condition. I call this third trajectory *past futures*.

For a while, it had seemed as if the presence of the climate crisis was restricted to the Global South. It unquestionably continues to be more readily discovered in exotic localities far removed from the centers of debate and policymaking. Thus, a small number of marginal spaces have spiraled into fame of global proportion, showcasing what is in store. The sinking state of Tuvalu, eroding remote Indigenous settlements along Alaska's shores, or the Sundarbans hugging the Bay of Bengal appear to capture some of the horrors associated with sea level rise (Marino 2013; Farbotko 2010; Farbotko and Lazrus 2012; Centre for Science and Environment 2012; Ghosh, Bose, and Bramhachari 2018). All these places inhabit what Michael Goldsmith (2015) calls a "big smallness" since their size, numbers, or geopolitical relevance stand in stark contrast to the role they seem to play in forecasting global futures. At first sight, the Sundarbans appear as an exception here, home to several million people, spread across hundreds of islands and peninsulas. Yet, by South Asian standards, the Sundarbans too might very well appear of a "big smallness" aggravated by their marginal status.

This tendency to discover climate change in exotic places is only partly explained, if at all, by issues of social vulnerability. It reflects a trend of reading, and subsequently heralding, entire global regions as zones of volatility at the hand of wrathful natures, rocked by geological and meteorological upheaval (Bankoff 2003). An obsession with treacherous environments—both seductive in their abundance and shocking with their violent outbursts—has catered to the anxieties of travel writing and the manly conduct of adventurers moving beyond the edges of the temperate West (U. P. Mukherjee 2013). It has also fed into broader patterns of rendering the Non-West Other as zones of chaos and despair, waiting to be pacified and brought to fruition at the hands of (post)colonial enterprises (Said 2014). Superior technological knowledge, "proper" work ethics, and draconian measures could then be legitimized, and desired, as a means of taming the wrathful Other, be it of a social or natural kind. Irrespective of the demise of the colonial project, these patterns continue to wield enormous power. And they make a reappearance in a different guise, I suggest, in climate change debates. Nature, or so they imply, seems to gather force in exotic localities far removed from the West. If debates on climate refugees are anything to go by, these localities are—once again—helpless and in dire need of assistance, be it in the form of infrastructure for drying or drowning lands, passports for citizens of sinking tropical nations, or aid for impoverished masses (Dalby 2005; Hartmann 2010; Piguet, Pécoud, and de Guchteneire 2011).

Furthermore, the reporting on lives in these new disaster zones is informed by a rhetoric of spectatorship. The plight of people already enveloped in sea level rise, desertification, or increasingly fierce storms is, I posit, made relevant as a warning, as a glimpse into what the future might hold for other places. For this purpose, those sticking to increasingly inhospitable lands or weathering devastating events narrate their troubles into cameras or are cited in glossy reports. Sundarbans islanders, alongside Pacific Islanders or Indigenous Alaskans, are put in the spotlight in order to give an insight into anticipated onslaughts, to see possible futures in the present. "Ghoramara is not just any island. It is symbolic of a problem that transcends local, regional and national boundaries. It is the actual face of global warming and climate change, the biggest problems facing the earth today," said Sunita Narain, arguably the most prominent environmentalist in India today (Niyogi 2009).

Ghoramara also features prominently in one of WWF-India's reports. In the introduction, it reads: "Their [Sundarbans islanders'] lives remind us of how precarious our existence is" (WWF-India 2010, 1). In attempting to direct attention to specific patterns of suffering, such portrayals mobilize people into being proxy witnesses for audiences far removed. Philosopher Hans Blumenberg (1997) explored what he understood to be a key metaphor in Western thought involving shipwreck and spectators at some length. To him, this figuration was noteworthy for reappearing in wildly different contexts and for emphasizing a not-so-subtle joy among spectators for being safe as they watch other people's misery. Media representation of climate change's proxy witnesses actualizes the metaphor. Again, spectators are witnessing other people's perils from afar, while also shuddering at the very possibility of meeting similar fates.

Within this ambiguous terrain, the Sundarbans take on a distinct role. Temporal inflections have an added twist. As I demonstrate, dreaded futures are both a thing of the past and an ongoing menace. Media accounts only momentarily dwell on what is currently happening, if at all, before quickly turning to islanders' present as a glimpse into the future. But, as we will see, islanders look back on decades of coastal onslaughts, assessing and evaluating *bhangon,* coastal erosion, by considering their past, present, and future. In turning to recollection of past and present practices of coming to terms with rough waters, this book unearths future possibilities as past conditions. I call attention to lives that are already being lived through what elsewhere appears as a rather dystopian vision.

Provincializing Climate Change

In unearthing future possibilities as past conditions, *Enduring Erosions* also contends with using the Sundarbans as a textbook case of climate change victimization. Film teams and professional photographers flock to the islands to produce truly haunting visual material. Over the years, rich documentation has emerged in the form of glossy pictures displayed in art galleries, in YouTube videos, or in scholarly and journalistic texts, all drawing attention to the sinking Sundarbans (see, e.g., Padma 2019; Cockburn 2010). Common to the vast majority is the conviction that what happens here is climate change and that the plight of islanders has something to say beyond local despair or ethical concerns. Images of sinking lands and collapsing houses, maps indicating the decreasing size of landmasses, and texts invoking the horrors of displacement are pitched as a wake-up call to the reality of climate change–induced sea level rise. To be sure, most outputs by artists, journalists, and scientists attest to the diversity of environmental changes that culminate in embankment collapse and shoreline erosions. But these texts and images appear to be organized around climate change, embodied by sea level rise. And there is reason to do that, simply because the Sundarbans today are a palpable zone of global warming's existence (Mortreux et al. 2018; Hazra 2012).

However, the climate change label can be a misnomer of sorts, and this book takes a more nuanced stance. It does so on three accounts. It first situates current suffering and challenges in their historical contexts, demonstrating that the Sundarbans have been a frontier zone all along, with their watery edges being the site of immense suffering. Second, it argues that the effects of global warming are but one class of anthropogenic interventions into the waterscape causing severe distress and a distinctly textured type of disaster. And third, it demonstrates that the anthropogenic dimensions of coastal erosion, including climate change, are very often cast aside or overshadowed by islanders dealing with more immediate issues. In seeing these concerns through, this book is—to update Dipesh Chakrabarty (2001)—provincializing climate change. If it holds true that climate change demands new histories and new forms of thinking the human, culminating in the notion of the Anthropocene, as Chakrabarty claims elsewhere (2009), these takes need to account for how climate change inserts into already volatile worlds. To call attention to the unevenness of the Anthropocene, to emphasize that the *anthropos* is not a unified actor but splintered across the lines of class, race, locality, or time, certainly remains an essential endeavor (Haraway et al. 2016; Bauer and Bhan 2018). In addition, climate change also requires being situated as an experience, as a lived reality amid a range of crises and environmental degradations making up life at the margins. Provincializing climate change, then, does not aim to put climate change in perspective or deny that global warming poses existential threats to the archipelago sandwiched between the Bengal floodplains and the bay. It is rather an attempt to situate climate change as process, imagery, and experience in the texture of everyday life on these islands in order to do justice to the complexities of environmental degradation. Yet, among multiple threats and degradations, climate change comes with unique traction. It stands to reason that the ubiquity of "climate change" not only overshadows the multifariousness of degradations but also comes as a convenient signifier to put blame elsewhere and thus negate any detrimental consequences of nearby anthropogenic interventions. To provincialize

climate change thus combines being attuned to local experiences and
tying them back to diverse transformations and interventions in time and
space. It considers responsibility to be close by and afar, further
anchoring the concerns of postcolonial studies in coming to terms with
planetary turmoil.

Plan of the Book

This book explores how people endure erosions in six substantial
chapters. Chapter 2 situates my ethnographic account by drawing out,
very briefly and necessarily sketchily, the contours of the social and material landscape of the Sundarbans and Bengal delta.

Chapter 3 explores the specific texture of coastal erosion. Following my friend Nur to the ruins of his former home, I account for a gap
between disaster policies and what counts as truly disastrous occurrences
along the coast. Official disaster response appears ill-suited to address
coastal erosion. As a chronic and dispersed dynamic that hardly ever
directly costs lives, coastal erosion remains well below the radar of disaster management. I take this to argue for a different understanding of
disasters. Coastal erosion, I suggest, is best understood as a disaster distributed across time and space. Short of serving as a flattened-out chronic
condition, it affects people and places differently. The chapter demonstrates how coastal erosion affects individuals and small groups during
times of endangerment, which can be mapped on seasonal rhythms and
disrepair. I thus shed light on the different temporal faces of erosion. It
can quietly linger over people across years and decades, as they find themselves at the watery edge, before dramatically bursting to the fore during
rather well-defined moments of risk. This leads me to argue that coastal
erosion is a disaster marked by a high degree of certainty—certainty about
when, where, and whom it will befall. Yet although coastal erosion is distributed, and as such spells out destruction in an individualized manner,
chapter 3 also traces collective dimensions. These emerge not only in
accruing large groups of displaced people across time and space, but also
apply to people living through seasonal rhythms and sharing narratives
of specific events.

Chapter 4 takes issues with mobility being considered as an exception. Exploring divergent historical trajectories that continue to shape
environmental relations, I show that islanders look back on deep histories of migration and mobility against the background of scarcity, conflicts, and disasters. Involuntary migration seems to have been the norm,

and spread out across long trajectories of what I call "circuits of displacement and emplacement" at—what used to be—the forest frontier.

In chapter 5, I explore resettlement. I demonstrate that rehabilitation used to be an achievement wrought by mobilizing favorable regulations within a climate of land redistribution, and put them to work in favor of environmentally displaced islanders. But resettlement was far from being a placid or homogenous affair, and I carve out two trajectories: from tense relations culminating in a death foretold in one to rather peaceful settlement operations in another. Considering these trajectories reveals the relevance of political and material contexts for actual resettlement experiences, and the salience of fine-grained analyses.

In chapter 6, I analyze embankments as vivid social sites. In a departure from the widely acknowledged fact that embankments are critical for life on the seafront, I take stock of a variety of embankments ranging from the simplest mounds to concrete seawalls. I read embankments as instruments of landscape transformation that instantiate state care or abandonment and profoundly shape everyday life. This leads me to consider the many roles invested in embankments. I show how embankments appear critical venues for income generation and for accessing scarce poverty-alleviation state funds, precisely in their frailty and disrepair. The vulnerability of coastal stretches and economic opportunities appear entangled in paradoxical ways. Below the level of authorized works, I find embankments to be fortified in unauthorized, ad hoc ways. I introduce the notion of minor infrastructures to account for such largely unacknowledged practices geared at upholding failing embankments. This involves unauthorized forms of fortifying embankment stretches deemed critical to businesses or of fixing frail embankments during surges, as well as everyday modalities of caring for homes and enabling human life in areas subject to tidal incursions.

Finally, chapter 7 analyzes ways of attributing meaning and securing a sense of hope on an eroding landmass. To islanders, the precise texture of the waterscape appears as a conundrum of engineering interventions, marine trade routes, playful currents or waves, and the very body of divinity. In demonstrating uncertainty in the attribution of blame and shifting trajectories of explanation, I draw attention to the emergence of what I call the other Ganga on these shores—the one that stands in a largely implicit but consequential opposition to the Ganga revered across India. I show how, in attempts to secure a future, islanders invest hopes in the protection of a hypermasculine god and similarly situated politicians able, or so it seems, to hold the complex muddy waters at bay.

Chapter 2

A Postcolonial
Archipelago

E rosion—counterbalanced at times by accretion, or by
the silting of rivers at others—underpins the dynamic
character of the Bengal delta. The Sundarbans, that
coastal archipelago serving as its delta front facing the sea, take this
dynamic character to the extreme. Here, rivers meet tides. The lands sandwiched in between shift, if left to their own devices, almost as quickly as
the waters do. But this is not "nature" alone, if such a thing were to exist,
nor are the flows rubbing against the lands only material. As any landscape, the Sundarbans are situated within larger force fields, and to
account for the dynamics at play here requires some mapping. This is
what this chapter sets out to do.

My attempt at mapping embraces the incomplete and the interwoven. I do not aim at a definitive or comprehensive account. Instead, I foreground a set of enduring flows enfolding the terrain now conveniently
subsumed as the Sundarbans. In this way, I will map out the Sundarbans
in time and space, highlighting some of the colonial and postcolonial
entanglements that sustain the texture of everyday life on the islands
sandwiched between some of Asia's most powerful rivers and the ocean.

So, what are these Sundarbans, and where are they?

Land of the Tides

As a first indicator of its multiplicity, the name Sundarbans holds a
layered etymology. Currently, it is most often translated as "the beautiful forest." Alternatively, it is also seen to be an appropriation of the term
"forest of the sundari trees" (*sundari ban*), a variety of mangrove, or of
the sea forest (*samudra ban*). Note also the plural *s* in Sundarbans, which
nicely highlights how this term, and its usage, is inflected by post/colo-
nial politics, since the plural marks an addition catering to the demands
of English grammar. It is redundant in Bengali.

One reason to think of the whole archipelago as the Sundarbans, including the rugged terrain made up of adjacent peninsulas and headlands, is the pervasive usage of the term as a label in governance regimes for the whole delta front. Another is the shared history of fairly recent settlement, and a very recent past, and present, of struggles with salty muds and their denizens. Decisive for me, however, is their shared reality as fleeting lands churned through by the tides. It matters little, in this regard, if the tides still enter land, as they do on islands free of embankments, or if the tides provide a looming threat held at bay only at great cost.

I find another name better suited to describe the region, one in circulation long before the denominator Sundarbans. It continues to be so on the ground and in fiction, but has lost traction regionally and globally. Bhatir Desh, the land of tides, pays heed to the critical role marine waters play in these unstable lands, while making room for the variety of environmental relations present—some of which may involve forests, others rivers or the sea, and others monoculture crops or fisheries.

Speaking of Bhatir Desh, the land of tides, also evokes how a majority of visitors continue to be equally fascinated and repelled by these impermanent lands, which bring the troubles of South Asian monsoon-crazed rivers to the boil, and the provocations both hold for modernist visions of neatly anchored, land-based forms of life. The repulsions combine aesthetics and governance.

Tides Pushing In from the South

Geographically speaking, the Sundarbans form—or at least are a significant part of—the front of the world's largest delta. Converging three of Asia's major rivers—the Ganga, Meghna, and Brahmaputra—it is continuously renewed and reshaped by astronomical amounts of water and sediments washed annually through the intricate meshwork of rivers and channels, and across the length and breadth of flooded areas, into lakes, ponds and wetland paddy fields teeming with life. Vast, and subject to a host of geomorphological, hydrological, biological, and anthropogenic influences, the delta is in itself hugely diverse. Differences stem from the overall growth of the delta—pushing it steadily toward the southeast, rendering the rivers on the western parameters increasingly defunct as they receive less and less sediment-laden freshwater and are, therefore, increasingly dominated by the tides.

Colonial administration was a watershed, quite literally, in terms of relating to excessive estuarine flows. Floods used to be an integral part of virtually all ecologies across the delta. Now they were a problem, and a menace to keep at bay. Efforts to quell them were never entirely successful, producing a number of severe repercussions. Disrupting the reach of flood waters onto land meant disrupting their growth and nourishment by sediments left behind. Left in the rivers, sediments would raise beds and choke rivers. Disrupting regular floods—diurnal or seasonal or both—dried lands, which would result in the compaction of ground layers as they dry and thus their sinking.

Vagaries of nature continue to frustrate aspirations inflected by high modernism. No single approach works in this terrain marked by intricate entanglements and fleeting currents. Much of the Sundarbans, and Sagar Island in particular, came to be known through the lens of disasters. Reading up on the island in gazetteers or travel writing, you learn of it through long lists of cyclones and surges, and the tales of obnoxious jungles packed with beasts. This is typical for the way the salty swamps were made to appear in public opinion, in a mix of fear and aesthetic disapproval of these lands so far removed from modernist visions of order, beauty, and promise. It only changed, partly, with a tourist obsession with exoticized jungles harboring startling wildlife in their monotonous appearance. At the same time, this take appears actualized in contemporary climate change debates.

Archipelagoes of Governance
The present-day Sundarbans present a matrix of partly overlapping, partly interlocking administrative units. They are divided between India and Bangladesh, with much of them making for a dynamic border zone notorious for smuggling, ambivalent citizenship claims, and piracy. Simultaneously, they are divided into a protected zone and buffer zones spread out across both national territories. Taken together, they form the transboundary Sundarbans UNESCO Biosphere Reserve.

The protected swamps are the world's largest uninterrupted mangrove forest, home to the largest population of royal Bengal tigers. The origins of the protected area, however, precede the fascination with big cats, and reach into a time when the government still paid head money for every shot tiger. After the wholesale clearance of woodlands, which amounted to 70 percent of the original forest being cut down between 1830 and 1873 (Dewan 2021a, 301; cf. Richards and Flint 1990), in 1878 the colonial government set aside substantial chunks of what was left in the form of reserved or protected forests (Sarkar 2010, 88–89). They felt this necessary in view of projected future demands of timber in Calcutta,[1] the then capital of India. Not just an example of the close entanglements of corporate interests, colonial governance, and forest protection—and how these gave rise to modern nature conservation—this also indicates entanglements between the coastal archipelago and the metropolis.

Today, India's Sundarbans fall into the district South 24 Parganas. The name stems from the twenty-four kingdoms the British East India Company took as a lease when they established their presence on the subcontinent, and invokes, to me, a combination of ruggedness and density that continues to haunt the territory. Its northern parts are rapidly urbanizing, and bleed into the metropolitan area of Kolkata. With the exception of a few fishing harbors and administrative centers, the south remains predominantly rural. Even so, the population is large, its density high, even for South Asian standards. According to the 2011 census, the most recent authoritative numbers available, the district population exceeds eight million people, with a little less than half of it rural.

In settled regions, mangroves have almost entirely given way to paddy fields, ponds, and houses, which, taken together, conjure up visions of rural well-being. With most homes adjacent to fields, villages spread out, rather than compacting into neighborhoods or lanes, which begins to explain why in areas badly affected by erosion people lose homes every year.

Scholars agree on the leveling effects of life on Bengal's forest frontier. Social hierarchies appear less pronounced than in other parts of the subcontinent. On Indian terrain, the otherwise rampant divisive politics of Hindu nationalism have so far made little headway. Similarly, the politics of caste seems less pronounced, while still being relevant for lifecycle events, such as marriage, or the solidarities underpinning how welfare distributions come to be routed or how career choices play out (Chandra, Heierstad, and Nielsen 2015).

The majority of the population is designated as low-caste Hindu (largely Mahishya caste) or Muslim (largely Sheikh), groups that have been noted for their mutuality (Nicholas 2003). These demographic trends also serve an indicator that the Sundarbans house large marginalized and vulnerable populations, people who often are "too poor to finance migration" (Black and Collyer 2014, 290).

---

1. In 2001, Calcutta was officially renamed as Kolkata. Across the following pages, I will use Calcutta when referring to the city prior 2001, reserving Kolkata for later years.

Party Politics and the Ghosts of Partition

Partition split the Sundarbans in two. Dividing British India into India and Pakistan drove a crooked line through the archipelago. But if the partition in the West was an upheaval, driving millions of refugees amid horrendous carnage in both directions across the new border dividing the Punjab, the partition in the East, dividing Bengal, turned out to be more of a gradual and less violent affair. Studies of the partition in the East concentrate on urban centers. On both sides of the border, the Sundarbans saw the departure and arrival of people fleeing conditions in their former homes turning increasingly caustic, because they appeared as strangers in their own land on grounds of religion. Since many of the absentee landlords were based in Calcutta, the partition cut them from their dominion in the eastern parts of the Sundarbans, leading to disarray in the provisioning of coastal protection as funds dried out (Dewan 2021b). In the western parts, in what was now independent India, refugees from the East gave a significant boost to artisanal fishing, and their descendants continue to be admired for their superior skills and the capital amassed in the form of trawlers and through the trading of fish.

In terms of party politics, the Sundarbans could be considered a backwater of sorts. With very few exceptions, such as the communist-led Tebhaga Movement of 1946–1947, championing the rights of sharecroppers, the region saw little of the contentious politics rocking Bengal and, later, the state of West Bengal (Dhanagare 1976). It never became a hotbed of further communist agitations, a trademark sign of politics in West Bengal since the struggle for independence, or of the mobilizations leading to the demise of communist rule in the state.

Political violence flared up, however, during the 1979 Marichjhapi massacre, when a large group of refugees from Bangladesh fled a state-sanctioned resettlement operation in parched central India and took refuge on an island in the Sundarbans National Park. Very soon, armed forces cordoned the area and killed, wounded, and ousted a still-unknown number of refugees (Jalais 2005; Mallick 1999). With the jungle having taken over the island again, and the victims or perpetrators never accounted for, the incident continues to haunt public imagination.

Similar can be said about another incident of violence in South Bengal. In 2007–2008, farmers clashed with armed forces over forced land acquisition for the benefit of a proposed Special Economy Zone in Nandigram, right across the Hugli River, in an area that many Sundarbans islanders consider their ancestral home. While this movement would eventually contribute to the ousting of the communist party from power, the agitation never took hold in the Sundarbans. It sent ripples, however, through rumors about the disposal of dead bodies on one of the sandbars of the delta.

Development at Sea

Long histories of state-backed, globally funded nature conservation projects notwithstanding, the Sundarbans have also been impacted by major industrial development. Even if most projects have come up in quickly urbanizing areas along major rivers, flanking the delta archipelago, concerns over detrimental impacts persist. Tightly interwoven through tidal flows and permeable mud lands, the delicate delta ecosystems—both forested and colonized—are also particularly susceptible to chemical contamination.

On the eastern perimeter, in Bangladesh, the recently opened Rampal Thermal Power Plant has caused much debate due to the urbanization and industrialization it is thought to attract. Concerns about toxic discharge and an increased risk of surges complete a complex picture.

Haldia, on the Hugli, clearly visible at night from Ghoramara and northern Sagar, is now home to East India's biggest hub of marine cargo handling and of petrochemical processing—the Haldia docks, part of the former Kolkata Port Trust.[2]

On the western perimeter of the Sundarbans, a number of further large-scale development projects have been stalled. These include turning the newly emerged island of Nayachar into a chemical factory complex, run by the Indonesian Salim Group on its own Special Economy Zone. Concerned scientists raised alarm about both the very likely toxic discharge and the vulnerability of the landmass to erosion. After the inauguration of the new chief minister, Mamata Banerjee—who rose to power also by helming a successful movement against forced land acquisition for large industrialization projects less than five kilometers away, in Nandigram on the Medinipur mainland—the plans were shelved.

A few years prior, the government had planned to turn the island of Jambudvip, straddling the sea a few kilometers south from Sagar, into

---

2. In 2020, the government renamed the Kolkata Port Trust into Syama Prasad Mokerjee Port, providing another instance of the trend, pervasive across contemporary India, to insert what appears to be Hindu tradition into the public. Since much of my research predated the renaming, and since the new name has gotten very little traction on the ground so far, in this book I retain the old name.

a tourist resort. After displacing fishermen who used to make camp there for their season and declaring the island restricted, the government axed these plans as well, to the relief of environmentalists and the frustration of fisherfolk still banned from entering.

The Gateway to India

It is ironic that the area now known as Sundarbans, once serving the British as an entry point to the subcontinent and for centuries bordering their first city, now appears relegated to the fringes. But this is not strictly true. The archipelago remains embedded in travels, and vibrant flows. Some align with the rhythm of the seasons and the tides, others do not. All intermesh the rugged, ephemeral lands with the surrounding world.

Media professionals continue to arrive, seeking to gather haunting material on the climate crises (Harms 2018). Development professionals come to experiment with adaptation (Cons 2018). Despite operating on their own terms, both inscribe the Sundarbans reiteratively as a climate change hotspot.

Many more travel here in the hope of seeing a tiger in its habitat. Since the royal Bengal tiger is considered not only a particularly charismatic big cat but also an emblem of Bengali prowess and the Indian state, it may not be exaggerating to claim that the tiger, tucked away in the remotest part of the forest, weaves the Sundarbans tightly into the Indian nation. Indira Gandhi's endorsement of Project Tiger has only further entrenched the protection of tiger habitat on the archipelago and provided for additional means to secure it through elaborate bureaucratic regimes (requiring permits to enter) and paramilitary forces (enforcing state control). In the long run, this fascination with tigers has also put the region on the list of sites of tourist consumption, increasingly also in terms of sun and sea tourism.

Beach tourism also adds to the vast numbers of pilgrims traveling to the Hindu shrine of Gangasagar, where, as I will explore in later chapters, Ganga is worshiped as she merges into the sea. Pilgrimage traffic culminates with an annual fair, India's second-largest congregation after the Kumbh Mela. When, on the days of the fair, the bus terminals at both ends of the road are awash with pilgrims heading to the annual Gangasagar Mela, the Sundarbans fold into Hindu India's sacred geography. The flows of cash, people, and ideas it involves become palpable.

These travels, of course, have their counterparts—the circuits of labor migration connecting one of West Bengal's poorest regions to employment opportunities near and far. In that respect, the Sundarbans are also known as Kolkata's maid, "Kolkatar jhi," essentially a pool of unskilled workers, both domestic and in construction (Ray and Qayum 2009; Jalais 2010b; Badiani and Safir 2009). The scope of labor migration, however, reaches beyond Kolkata, with many engaging in seasonal, circulatory, or permanent labor migration on a massive scale across India, and occasionally the Persian Gulf. Often, the tourist industry paves the way: islanders meet middlemen or future employers as they visit forests, beaches, or shrines, made possible by significant tourism infrastructure. The vast majority of Sundarbans households, whether adjacent to the reserve or on a settled island farther away, have people sent out, or are waiting to do so, and depend on the remittances, which—paradoxically—make possible the lives of farmers and fisherfolk.

An even darker undercurrent to this is the trafficking of brides from here to North India, where there is a dearth of marriageable women due to the widespread abortion of female fetuses. The demand is so high that people in North India frequently skip demanding dowry, and are even willing to pay fees to middlemen, thus enabling the marriage of daughters to swaths of the coastal poor unable to cough up the otherwise necessary dowry.

Perennial Migrations

Migration—as elsewhere—is not a recent phenomenon. The Sundarbans never were inert or empty, nor were their people ever standing still. Settlers from today's Myanmar, the notorious Arakanese; early Portuguese settlements; or the port of Hindu kings shrouded in mystery all stand testimony to that (Sarkar 2010).

When, in the nineteenth century, colonial authorities began pushing for settlement of the mangrove swamps, investors and state agencies tapped into poor people's reliance on migration as a strategy to counter rural distress. By then, poverty was rampant across southern Bengal. Heavy taxation, the concentration of holdings by a new class of rural gentry, and the ousting of local industries by British manufacturing culminated in rural stagnation (S. Bose 1993) had rendered populations footloose and ready to take risks elsewhere. Complementing a better-researched stream of paupers into urban centers (Chakrabarty 2000), another trajectory took vast numbers of people into the coastal swamps.

Recruited by investors holding leases toward large estates, rural poor began settling the swamp by axing trees, erecting embankments, and building houses. Salt manufacturing gave way to rice paddies, with investors hoping for profit through chains of subleases and rents, and the state eyeing a windfall in taxes. Given the extractive nature of endeavors, and the complicated distribution of responsibilities along chains of subleases, provisions for workers were meager, and life was harsh. A number of estates failed, as the investments needed surged, and profits were below expectations. Others were wiped from the face of the earth by recurring storms and surges. Poor coastal soils and frequent disasters, not to mention tiger attacks or vector-borne diseases, frustrated colonial ambitions to transform the archipelago into a seat of plenty and to look at it as an "Aladdin's cave" (Lahiri 1936, 39) waiting to be unlocked. While the state continued to invest in the vision of turning wastelands into productive plots, and of warding off obnoxious wilderness, the Sundarbans became the problem child they remain to this day.

In contrast to many other parts of deltaic Bengal, a class of rural gentry never fully emerged. Large landowners remained absent, and the chains of subleasing lands were frequently reshuffled, which, together with meager agricultural return on unaccommodating soil, blocked economic divisions from emerging as pronounced as they were elsewhere.

Individual holdings remained comparatively small, and the communist-led land reform, which aimed at distributing large holdings and state-owned lands for the benefits of landless farmers, made little impact (but was taken up as a means of achieving rehabilitation of environmentally displaced islanders, as I show in chapter 5).

Even if aspirations faltered, the colonial state managed to implement a regime of rural living based on private property arranged in neatly distinct plots. It catered to the demands of agriculture, gradually discouraging commons and other forms of collective control over natural resources. For all these purposes, embankments became critical. They were means of turning amorphous tidal swamps into territory—that is, of carving out stable terrain, having property relations fixed, and subjecting all to state control.

In this view, embankments did not solely keep salty waters out, they were also seen to guarantee the legibility of the land, pinpoint assets, and conveniently allocate plots to people, while also asserting some degree of state protection. In short, embankments were critical instruments of biopolitics on the coastal margins. Over the years, control over embankments and the responsibility to maintain has shifted frequently—first, from absentee estate owners to the government, before being pushed from one department to another. Presently it is divided between the Department of Irrigation and Waterways and the Sundarbans Development Board. On the ground, embankments quickly became vehicles to assess actual investments by estate holders or state agencies, as well as means of contesting the state and officeholders. As a consequence, the status of the embankments remains the focus of heated discussions, frequently serving as a gauge of current state provisions.

Chapter 3

Hidden in Plain Sight

Coastal Erosion as Distributed Disaster

The function of the building—an unassuming three stories—was easy to miss. My friend Nur showed me it when we visited the village he originally came from, Botkhali. By then I had known him for a while. We had spent afternoons talking on his veranda, enjoying the steady breeze that came as a bonus with living in such a precarious position, right on the outer embankment of Gangasagar Colony. He came late to the colony, together with eleven other families, all dwelling in close proximity to one another on the ring embankment that shields the colony from the tides. Not long after we first met, we decided to visit his former home, or what was left of it, on the rapidly eroding southwestern tip of the island. Our visit was another exercise in relearning the landscape. Back then, the village was awash with the activities of a famous international NGO. The NGO had identified Botkhali as one of the worst-affected disaster zones in the state and chosen it as a site for a project aimed at improving adaptation and resilience to coastal disasters. Over the years, as I got myself acquainted with staff at all levels, I made a habit of inquiring how its work fared. I listened to villagers evaluating what had been done and traced remnants in the physical landscape.

The project was ambiguous and multidirectional. NGO workers were acutely aware of the enigma of erosion. Collaborating with university-based scientists, they produced publications on coastal erosions geared at instigating policy debate (e.g., Ramakrishna Mission Lokashiksha Parishad 2009). At the same time, however, the practice of development assistance in and around the headquarters related only to classic disasters of the more sudden and spectacular kind.

Seminars were held, familiarizing villagers with first-response actions during storms and surges. They had the chance to try on life jackets and learn about rescue techniques that certainly would be helpful when disaster struck again. This was fun, villagers said. Educative plays were shown, music recordings were distributed, all instructing villagers about the nuts and bolts of disaster preparedness. The messages were clear: be prepared, listen and pay heed to forecasts, help one another during surges. Murals advised people to stock packaged food, potable water, medicine, and wood during high-risk months, or to boil any drinking water after surges had struck in order to avoid the spreading of disease.

The Lost Cyclone Shelter

To facilitate these and further activities, the NGO took up office in a village building sandwiched between the remains of Botkhali and the main temple of adjacent Manzabazar, in one of the very few multistoried buildings in the area. According to the standards at work by the time I began researching on Sagar, it certainly embodied sturdiness and signaled modernity. Built from brick and concrete, it promised safety from storms and any ensuing surges, by virtue of its ground floor sitting up above the foundations. For the duration of its operation, the building's upper floor served as an office space and outreach center. Workers invited villagers to use their headquarters as a cyclone shelter in the event of impending storms. People in Botkhali appreciated these efforts. But to me its fate became emblematic of the vicissitudes of dealing with, and thinking, disasters on these shores.

The murals were fading over the years. Their messages seemed somewhat off the mark. The area had certainly been badly affected by storms throughout its settlements. Indeed, Sagar Island as a whole appears in modern history through the lens of storms and surges, as a glance at revenue reports shows (see, for instance, Ascoli 1921, 118). However, villagers insisted that this was not the most pressing problem. Echoing others, Nur repeatedly stressed that the disaster-related activities of the NGO or of state agencies were all fair and good but did not really help them to deal with what they saw as the most pressing issue locally—the erosion of village land by brackish waters.

A few decades ago, Botkhali sat a comfortable distance from the sea. Villagers had always enjoyed ties to the sea, as a small channel, a *khal,* ran through the village, connecting it to seaborne trade, traffic, and fishing grounds offshore. In fact, folklore has it that hunters used the channel to enter the swampy jungle and land their boats (Maiti 1994). It used to be secluded by dense mangrove forests. A comprehensive settlement, homesteads, places of worship, and rows of embankments and fields acted like buffers. All this is gone. Tidal waters have plucked away everything that stood between Botkhali and the sea. Two large and prosperous villages have been pulverized by the encroaching sea, their remnants washed away and populations driven out.

The NGO's project did not have the means to engage with erosion, instead relying for all practical purposes on the conceptual apparatus and developmental toolbox of disaster emergency. In attempting to implement coastal resilience, ironically, erosions were well below the radar of this and all other projects in the region.

Adding to the frustration of villagers, this project's material traces withered and were literally swallowed by the sea. The initiative followed the usual short lifecycle of development projects. Once the project cycle was completed and the NGO stopped working here, the fate of the storm shelter became uncertain. With rental agreements terminated, it was unclear whether villagers would be able to use it in the case of impeding storms. What is more, shortly after the NGO had wrapped up its activities, the building itself became subject to erosions. With foundations eroding, it eventually collapsed, the remains washed away. While no one was hurt, the cyclone shelter has become a thing of the past. It would be wrong to read this as an example of inefficient or faulty engineering—as if the shelter would have needed a safer locality (removed from the shore) or should have been of a more robust construction type (better enforced). Both measures might have extended the duration of the building's existence. Yet given the rampant incursions by brackish waters, the dismantling of the former shelter was only a matter of time. It became another instance of the shortcomings of classical disaster management and humanitarian assistance in dealing with coastal erosions.

The problem, of course, is not one project nor one NGO. It is endemic. Its most recent avatar is the shiny new public cyclone shelter opened in 2018 only a short walk away. Situated in a revamped high school building, enforced by concrete, and equipped with a ramp, this shelter illuminates the continuing prioritization of means and tools of disaster management, concentrating funds on flagship infrastructure for extreme weather events. It stands to reason that this flagship building will be washed away by the waters in due time, much like its predecessor. For the time being, it articulates the rift between signature investments and the sorry state of collapse-prone embankments on much of the outer parameters.

Contrary to the emphasis of state or NGO-driven disaster management policies on extreme events, the normalized shifts of shores spell out disasters to coastal dwellers caught on the edge. Their terror rests precisely in being small-scale and contained, in being hardly discernable yet unstoppable. This chapter engages the paradoxical nature of coastal erosion marked, as it is, by mundane, barely visible transformations and extraordinary experiences, culminating in loss and despair. It uncovers coastal erosion as a phenomenon that is distributed and hidden in minor forms of ruination, involving large populations.

The following sections unpack the ways coastal erosion impacts the lives of islanders across time and space. My aim is to develop a conceptual apparatus apposite to tracing, and thinking, disastrous dimensions of coastal erosion. I demonstrate that erosion is currently the most pressing environmental hazard and has been for at least a few decades. To islanders, this is the disaster, dwarfing all other hazards. Yet on all practical accounts, coastal erosion remains below the radar of humanitarian interventions and disaster management. I maintain that this paradox refers to the very overflowing, and flooding, of neat social, spatial, and temporal parameters perused by disaster thinking. I am referring to noticeability, to what matters and enforces orchestrated action among state or humanitarian actors and distributed publics witnessing disasters from afar. Building on that, this chapter reveals coastal erosion to be a disaster of peculiar texture. I offer the concept of distributed disasters to capture its specificity. The notion highlights that coastal erosion affords experiences, demands navigations or coping mechanisms, and enables modes of collectivization in terms strikingly different from what is typically understood as disaster.

Firstly, to think of coastal erosion as a distributed disaster foregrounds its complex temporal patterns. Coastlines disappear and landmasses shrink not in an instant but in drawn-out processes. This involves miniscule, almost invisible damages, marked by "non-events" (Povinelli 2011), that come interspersed with, or culminate into, eventful aggravations. The language of slow-onset or rapid-onset disasters is of little help to account for the rhythms and open-endedness of erosions.

Secondly, the concept of distributed disasters invites us to think erosion as an individualizing process that affords specific socialities. Amid the individualized troubles of lost harvests or sunken homes, I uncover collective experiences as they emerge over time and space, and through one particular story. Socialities emerge in ephemeral, mobile slums housing freshly displaced islanders as well as in resettlement colonies, giving contours to the experience of environmental displacement.

Thirdly, this chapter calls attention to the spatial distribution of erosion across, for instance, delta-front archipelagoes. Here, pockets of simultaneously affected people are spread out across far-flung corners.

In framing coastal erosion as belonging to a specific type of disaster, somewhere between event and process, I call attention to a middle ground between disaster and ordinary life that informs life on a warming planet. It involves living in, and through, extended conditions of criticality interspersed with recurring periods of onslaught and destruction.

Coastal erosion is underpinned by what I dub deafening certainty. As many interlocutors revealed, it is not so much a question of whether specific villages and individual shorelines will disappear, but rather of when. In contrast to weather extremes that are often uncertain until the very last moment, or to lives impacted by chronic, hardly detectable toxicity, certainty appears to be writ large in the experience of coastal erosion. I call attention to a certainty of being in the way of encroaching waters and shifting shorelines. If the disaster is distributed rather than a slow-onset or rapid-onset event, its uneventfulness poses methodological challenges. After all, it does not affect social wholes. Nor does it sit in places, to update Basso (1996). It instead moves in on dispersed zones, holds them in its grip, and lurks in the interstices of everyday life. Likewise, at no point do groups attribute meaning or blame in a collective process of coming to terms with the situation, typical of slow- or rapid-onset disasters (see, e.g., Hoffman 2002; Button 2014; Sökefeld 2012). This demands apposite methods. Therefore, this chapter abstains from fully engaging with how erosion is lived through as it hits individual banks or embankments—this is reserved for my later chapter on the politics of coastal protection infrastructures. I explore coastal erosion as distributed disaster of deafening certainty by following Nur and his fellow islanders, tracing steps along this shifting shoreline and delving into narrations of pasts. Attending to how people remember their pasts reveals the texture of the disaster and offers glimpses into lives continuing amid the looming threat of future erosions.

What Is the Disaster and When?

Nur first showed me the building shortly after Cyclone Aila struck India's West Bengal and Bangladesh in May 2009. To many, Aila became a watershed moment. Climate change had arrived, cropping up in vernacular debates. Previously cyclones were referred to by their years, but Aila was the first cyclone to be given its own name. Displacement from the Sundarbans to the fringes of the city suddenly had both a reason and a face. Analysts claim rightly that while this storm took only a few lives, it had a lasting imprint on rural livelihoods across the Bengal delta. Its timing spelled havoc, and the long-term effects of saline flooding are pressing still (Islam and Hasan 2016; Mukhopadhyay 2009; Sundarban-basir Sathe 2010).

For better or worse, I was not on the island as the storm struck, nor was I a direct witness to cyclones Bulbul and Amphan, which wrecked parts of the island in 2019 and 2020 (J. Basu 2020; DTE Staff 2019). I was surprised by Aila in the early days of my research some two hundred miles inland. Along with my family, I was holing up in a concrete house. Incessant rain and winds tore at roofs and walls over the course of a few days, joined by several ant colonies, which had pushed through cracks in window frames in order to escape the gusts. As soon as I could, I returned to Sagar. Colony Para was safe. All embankments had withstood the wind and the waves. Trees were bent over, and crops bruised, but nothing seriously troubling had happened. Nothing to worry about, no inconveniences caused, one resident of Colony Para told me smilingly as he continued fixing his fence. Business as usual, or so he implied. Merely a storm, and in weathering storms they were experts.

With this storm freshly imprinted in her minds, Gouri, an elderly woman residing in Colony Para's longest-settled parts, later told me that storms are, well, just storms. This was a view that resurfaced throughout my research here. Trees fall, tiles fly away, and livestock gets washed away, but all of this is repairable, Gouri exclaimed. Livestock can be raised again, trees grow back, and roofs can be fixed. It takes effort and time, but it is not difficult. It is something that people are accustomed to. High winds occur regularly, demanding efforts to build back and to recover, and in some ways, they feed into a series of mishaps and setbacks that is life. Local societies were so well adapted that hazards simply did not, and could not, morph into full-blown disasters—akin to resilient "cultures of disaster" to be found in the Philippines (Bankoff 2003).

This is not to say, of course, that all was rosy. Some had terrible stories to tell of Aila and other storms. There were those bereft of family networks, like left-behind, childless Lakshmi Jana, who saw her roof collapse, and lacked the knowledge to muster resources and strength to build it back.

There were also those who had gone through storms in particularly vulnerable circumstances. Following storm warnings, people take shelter in concrete buildings or proper cyclone shelters, if within reach. Rumor has it that thieves would skim through deserted neighborhoods, robbing people of their belongings. Thus, in many cases, men stay back to guard their houses. Feeding rumors of higher male mortality during storms (but see for contrary evidence Doocy et al. 2013; Paul 2010), men report of the horrors of staying within shaking walls and in reach of whipped-up estuaries. Another trope of gendered vulnerability is that of suffering storms at sea. Since the main fishing season coincides with that of storms, almost all fishermen I spoke to had been through this. Some had paid no heed to storm warnings, generally putting little trust in the Indian forecasts compared with the Bangladeshi counterparts. Others had simply been caught out by high winds, too far out to make it to land in time (Gupta and Sharma 2009). They spoke of nearly capsizing, of broken-down engines, and, in extreme cases, of survival by clinging on to wood. All knew of colleagues and friends who had not returned. But in fact, all the terror and fright only fed into the common narrative: on land, a storm is just a storm. On land, there is firm ground to hold on to, and people are never really out of sight.

Coastal erosion is different. Once land is washed into the bay, there is nothing left to recover. To islanders, therefore, *bhangon*—the colloquial Bengali term for erosion, which translates literally as "the breaking"—counts as the disaster. It breaks away the foundations of terrestrial life-forms and, more specifically, the means to strive for agriculturally inflected visions of prosperity condensed in the alluring picture of golden rice fields and productive freshwater ponds. Such images are evoked most famously perhaps in Rabindranath Tagore's ode "My Golden Bengal," the national anthem of Bangladesh.

Prior to Aila, two particular storms were stamped into memory, the 1985 cyclone and, most vividly, the 1942 Medinipur cyclone, locally referred to simply as the Red Flood (*lal bonya*). Some understand the latter to refer to the reddish skin color of the then ruling colonial overlords, highlighting its occurrence in the late days of British India. Others contend it stems from the surreal occurrence of waters turning red due to their severe salinity. Either way, the flood is remembered as an instance of the temporary negation of land and fixity. It is here that the flood memories resonate with narrations of storm at sea and, more broadly, with coastal erosion. Depictions of drowned lands and submerged lives dictate these interpretations rather than storms and surges. Over tea, my friend Anil Patro, who had raised his children and grandchildren in Colony Para, explained, "The whole country [*desh*] was covered with waters and people were floating about." And his neighbor, who had joined our conversation, extended, "What was land once had become water." A number of my interlocutors told elaborate stories of how their ancestors had survived the deluge, avoiding "becoming fish," as another resident put it, by letting go of houses and trees. On not becoming fish, people became flotsam, holding tightly onto debris and trees, then floating with the currents. Some, miraculously, came out rich. In fact, the growth and development of the now-bustling port town of Haldia across the river is frequently considered to be an outcome of treasures dumped here by this flood. Many more, my friends insisted, were turned into paupers by the floods. Anil Mali vividly remembers his grandfather telling how the flood stripped him naked, barely alive and with nowhere to return to. Beyond embodied suffering and existential fear, this flood is, in other words, remembered as a reordering of property relations. Some were thrown into poverty, while one town was quite literally awash with riches. These are stories of the vast powers water holds. The way this flood appears in narrations echoes contemporary experiences of coastal erosion. Both feature the submergence of land, the liquification of fixed terrain, an implosion of clear distinctions between fluid and solid, and the reshuffling of property.

Beginning with Aila, all storms since have tragically hit Sagar on its fastest-eroding shorelines on the southeastern corner. Here the destruction was most severe, leading to permanent outmigration by serving the final blow to people who already knew that their time to move had come. The surge had rolled over the embankment, washing out the frail line of defense, flooding the zone behind and compounding erosion. To the people of Botkhali, later cyclones were more of the same. Aila was the start of a period of intensified coastal erosions, and now, more than ten years later, it serves as a timestamp against which to evaluate the brackish water incursions occurring before and ever since. Thus, even if the cyclone also appears here as a moment of catastrophe, the real disaster is something else. Meanwhile, islanders experience endemic and normalized processes as extraordinary and disastrous.

Situating Distributed Disasters

The discipline of disaster studies largely deals with events of gargantuan proportions. It is the study, or so it appears, of tsunamis drawing close; of earthquakes razing cities; of hurricanes literally carving themselves into landscapes, leaving trails of destruction. Disasters appear as events erupting into lifeworlds, shattering their very foundations and demanding responses both short- and long-term. Entire populations appear to hold their breath, sustaining direct injuries or witnessing in awe from afar. Thus conceived, disasters involve massive material destruction and economic loss, with wide media coverage. Let's take sociology of disasters as an example. Enrico Quarantelli (2005), a key figure in the field, insists his work is solely concerned with nonroutine events, differentiating between emergencies, disasters, and catastrophes. All are sudden and unexpected events that differ with respect to the degree of devastation and the afflicted society's capacity to manage. While specialists handle emergencies effectively by following their routines and procedures, catastrophes spell out a breakdown or utter irrelevance of societal response in the face of dramatic destructions. This way of thinking disaster fails coastal erosion, I argue, as it does a range of other degradations.

Scholars in the field of disaster studies, including Quarantelli, have long questioned framing disasters merely as events. To think disasters only as events means sleepwalking through the processes in which disasters build up beforehand, and through the processes of coming to terms with them politically, socially, or psychologically in the aftermath. Yet even with embracing drawn-out temporalities, these models still hinge on an event, a shock, or a turbulence.

Echoing German novelist Max Frisch's dictum that nature knows no disasters, scholars agree today that any given disaster emerges only at the intersection of circumstance and external trigger (see, e.g., Zaman 1999; Pelling 2003; Luig 2012; Oliver-Smith 1999b). In an influential study, Ben Wisner and colleagues (2004) distinguish between root causes and more immediate circumstances that render specific spaces or groups vulnerable to specific triggers. In this view, disasters have deep histories and coalesce with processes such as capitalism or gender disparity. They are, in the words of Anthony Oliver-Smith (1996, 314), "functions of an ongoing order, of this order's structure of human-environment relations, and of the larger framework of historical and structural processes, such as colonialism and underdevelopment, that have shaped these phenomena." Hence, disasters are conceived of as culminations, as explosive disruptions of the everyday, arising when shocks ignite unsafe conditions and as social vulnerability meets a trigger.

Triggers can also be the result of a process. Famines, for instance, rise out of a fusion of embedded historical context and external triggers such as droughts (Watts 1983), decision-making in war economies (A. Sen 1981), or a combination of factors (Tauger 2003). In other words, slow-onset disasters still rely on a language of events and eventfulness. Such a language speaks well, of course, to the drama, sensual richness, and

deep-cutting ethical concern such cumulative events embody. But studies have also shown that the notion of vulnerability to a given event
needs to be paired with a concern for how disasters unfold over time
(Hastrup 2011) and space (Nancy 2014). In the Lisbon earthquake in
1755, for instance, it was not the quake itself that caused most fatalities,
but ensuing fires and diseases (Molesky 2015). Furthermore, the earthquake left its lasting imprint on European debates over years of trying to
understand its implications (Neiman 2002). As for India's Gujarat earthquake in 2003, while the direct death toll may have been higher from the
earthquake, people on the ground saw their lives gripped by this disaster
for prolonged periods after the earth shook (Simpson 2014). In the long
aftermath, the experience of an angry earth involves a political and moral
realignment of society, identifying wrongdoing as its cause and using it as
a means of strengthening control (Schlehe 2010; Simpson 2011; Dove
2010). That is to say that even conventional disasters are in the process
of becoming for prolonged periods of time, as differently situated actors
react to and navigate through them.

Highly publicized disasters send ripples across the globe, changing
their shape through this attention and the ensuing debates. Many disasters feed into long-standing perceptions of certain regions as problem
children in need of help, reinforcing forms of Orientalism in the language
of disaster victimization (Bankoff 2003; Ethridge 2006; Watts 1983).
Humanitarian response—mobilized or bolstered as images of despair and
destruction circulate globally—inserts itself into the aftermath (Simpson
and Alwia 2008; Miller and Bunnell 2011; Hoffman 1999), considerably altering disaster becomings.

The aftermaths of technology disasters stretch across generations.
When the gas repository exploded in Union Carbide's pesticide factory
in Bhopal some forty years ago, the poisonous cloud did not only kill
thousands *that night*—as survivors put it, it entered the earth, plants, and
water of the city, eventually reaching the flesh and bones of its populations. For many of those who survived *that night,* but who could not and
still cannot flee, life turned miserable. Living after the cloud, the everyday became toxic and might be understood as a persistent disaster (V.
Das 1996; Rajan 2001). Thus, Bhopal now stands for both the aftermath
of an event and the still-unfolding presence of disaster. Anthropologist
Kim Fortun (2000) demonstrates how the state tried to enforce a sense
of closure that amounted to a grave simplification of the toxic and continuously unfolding nature of the event. Although the enforcement of a
settlement with the victims clearly followed political interests, it also

points to the difficulties in addressing unbounded processes and making them liable to claims for justice.

The troubles continuing to affect Bhopal are mirrored in a range of landscapes marked by encroaching ruinations and elusive toxicity (cf. Stoler 2008)—be it around nuclear testing sites (see Kuletz 2001), at industrial compounds (Auyero and Swistun 2009; Kane 2012; Tironi 2018), or within reach of large-scale development projects (Glantz 1999). While in Bhopal and elsewhere suffering can be pinpointed to a certain event and a group of actors (*that* night, *the* company), in many other cases, time, responsibility, and, indeed, the hazard are buried in what Auyero and Swistun (2009, 6) call "toxic uncertainty." Whether disturbance and disrepair can be tied to an event or not informs localized ways of living through and of making sense of toxicity.

Today's global environmental degradations, including the climate crisis and biodiversity loss, foreground another figuration of disaster. Scientists, activists, and journalists regularly employ notions of impending collapse, of massive disruptions and gargantuan disasters, to alert their readership, while also feeding into long-standing "horrified fascination to catastrophes" (Asad 2007, 73). At the same time, they highlight the quotidian effects of global warming or biodiversity collapse, deeming them a disaster, albeit ones operating on an open-ended temporal frame (Crate and Nuttall 2016; Baer and Singer 2014). Among processes of such gargantuan proportions, there seems to be no single event, nor is its wake identifiable. Prime for analytical purposes in this age of fallout (Masco 2015), toxins float through the atmosphere and oceans, intersecting with one another and smoothly crossing borderlines, bodies, and their environment (Murphy 2015; Liboiron, Tironi, and Calvillo 2018; Davies 2018). Repercussions of classical disasters, such as industrial accidents, overlap with intentional destructions, such as nuclear bombs, and with the destructive consequences of normalized practices, such as the use of chemical fertilizers or excessive greenhouse gas emissions. The search for straightforward causal relations, the attribution of blame, and the language of events and their wake all seem to falter.

Grappling with these complex temporalities and uneventful horrors calls for novel concepts. This book contributes to this search, offering the concept of distributed disasters. I draw on theorizations of quotidian crises, of everyday life left uncertain and unsafe amid shaken, poisoned worlds. Writing on the Aral Sea, Michael Glantz (1999) notes "creeping environmental problems" that cause a material decline of the water body and render biotic lives ultimately impossible. Enlarging the scope and turning to sweeping degradations, Rob Nixon (2011, 2) speaks of slow violence. He notes in a programmatic passage:

> By slow violence I mean a violence that occurs gradually and out of sight, a violence of delayed destruction that is dispersed across time and space, an attritional violence that is typically not viewed as violence at all. Violence is customarily conceived as an event or action that is immediate in time, explosive and spectacular in space, and as erupting into instant sensational visibility. We need, I believe, to engage a different kind of violence, a violence that is neither spectacular nor instantaneous, but rather incremental and accretive, its calamitous repercussions playing out across a range of temporal scales. In so doing, we also need to engage the representational, narrative, and strategic challenges posed by the relative invisibility of slow violence. Climate change, . . . deforestation, the radioactive aftermaths of wars, acidifying oceans, and a host of other slowly unfolding environmental catastrophes present formidable representational obstacles that can hinder our efforts to mobilize and act decisively.

His take is illuminating in theorizing the environment as a domain of specific forms of violence, echoing earlier work on structural violence (Farmer 1996, 2004) and social suffering (Kleinman, Das, and Lock 1996). If, as political ecology writing insists, the environment is accessible, and conceivable, only through webs of power; if the entitlements to nature's bounties and the possibilities to flee nature's wrath are always uneven; if nature flows from social practice; and if futures are lost due to anthropogenic degradations (Bryant and Bailey 1997; Escobar 1999; Peet, Robbins, and Watts 2011), then it is apt to conceptualize certain environmental transformations or states as a form of violence (see also Hartmann and Boyce 1983; Watts 1983).

Nixon's take also aids the rethinking of catastrophe from event to a process spread out across time and space. It sheds light on forms of chronic crises grounded in anthropogenic degradations. Chronic crises, as Vigh (2008) notes, alert us to the fact that marginalized lifeworlds require the navigation of critical states not as an exception but as normalized condition.

Recent scholarship on life and loss in the Sundarbans mobilizes such a framework. Aditya Ghosh (2017) sees people in the Sundarbans going through, and living with, everyday disasters. In contrast to Gregory Button (2014), who theorizes everyday disasters as the mundane openings and possibilities that allow prominent disasters to emerge, echoing discourse on vulnerability, Ghosh turns to quotidian processes that need not grow into a disaster proper. In this view, disaster never ends, but lurks everywhere and serves as a condition for lifeworlds and mundane decision-making. Amites Mukhopadhyay (2016, 12), on the other hand, turns, as I have been doing, scholarly attention to events with the "uneventful and imperceptible process of land erosion and embankment collapse." To him, disasters in the Sundarbans are a "frequent, mundane and everyday phenomenon" (13), arising in a context of "power and development apathy" (14).

Taken together, these accounts expertly illustrate the unbound temporalities of disaster along these threatened coasts and highlight how multiple crises can enfold the everyday. They resonate with Roitman's (2013) notion of anti-crisis, in which she critiques the concept of crisis, which seems to suggest a before and after, when conditions have in fact been ongoing. But they fall short, I suggest, when it comes to retaining the still-eventful nature of embankment collapse and individual displacement that comes to pass as cyclone shelters collapse or as people abandon homes making way for waves. In other words, to think disasters as everyday occurrences amounts to a dilution that leaves unaccounted for and undertheorized how exactly erosion impacts and is being dealt with by people on these watery edges. It fails to account for the specific texture of this disaster, involving, as I suggest, periods of quiet yet chronic erosion sporadically catalyzed by eventful outbursts. After all, as I demonstrate in this book, to live with coastal erosions on these shores involves being encroached upon by brackish waters, being flooded, moving out only to be caught up again by the same waters. Coastal erosion, I argue in this chapter, sits uneasily—as a process and experience—between chronic crises and what is commonly framed as a disaster. It forms a condition for prolonged periods of time, framing everyday practice and holding islanders in its grip. But within this, certain moments of a heightened drama stick out.

What is more, populations receding with the coast are frequently revisited by disasters, as waters keep on encroaching upon their sites of refuge after initial displacements. I frame these constellations as distributed disasters. In doing so, I specifically emphasize their distribution across time and space, and their patchy texture, involving times of respite followed by times of intense onslaught. The analysis of erosion needs to take these complicated patterns of effect and temporalities into account. Not only for the sake of theorizing disasters along sinking coasts or, by extension, in worlds losing their hospitality. Nor only to understand what Povinelli (2011, 13) calls "quasi-events," by which she means "forms of suffering and dying, enduring and expiring, that are ordinary, chronic, and cruddy rather than catastrophic, crisis-laden, and sublime." But also in order to work toward more apt societal responses. As much as climate change, and interrelated forms of environmental degradations, force us to rethink notions of history or nature (Chakrabarty 2009), they also demand a reconsideration of the temporalities of disasters, and, ultimately, the entire concept itself.

In the next sections, I account for the complex texture of coastal erosion. I begin by turning to a moment of heightened crisis, an inflection in drawn-out processes. Such moments of crisis serve as an entry point to assess the predicament at large. Focusing on this recurring moment also highlights, as I will show, the seasonality of erosions and their ongoing presence as people remember past incidents of incursion, loss and flight, and how they relate to times of threat as a protocollective.

The Dangers of the Full Moon

Before dawn, the waters were pushing in full force. Today was the peak of one of the particularly rough spring tides rocking the rainy season. The estuary was roaring as if in agony. Drawn into action by the moon, water levels had risen by more than ten meters, dramatically transforming the contours of the island. Waves and currents feverishly besieged all its outer ring embankments, gushing into any opening, however small. Tiny streams swelled into unruly rivers, drowning plants and elevations beyond the dyke. Fishing boats that a few hours ago lay stranded in mud now were afloat, tugging relentlessly at their anchors, while all sorts of marine creatures found themselves pushed deeply inland, along the channels, and washed up on embankments. As all tides, these too were involved in rewriting the limits of the land, the precise extension of the estuary's sway.

On drier parts of the coast, many people were wide awake. Some lay sleepless in their huts, listening to the gurgling and roaring, hoping for the embankments to withstand the onslaught. Others rushed along the embankments, looking out for weak spots and joining attempts to reenforce the meek wooden and muddy structures wherever needed.

The onslaught came close to toppling Gangasagar Colony's outer embankment. The settlement, as large parts of the island, showed its most fragile face: built on land below mean sea level, enclosed by mere muddy walls, besieged and bullied by waters now looking down onto farmlands and houses. It hurt to see its farmlands and houses being so deep below water level, protected only by weak dikes, so obviously on the brink of being washed away. Along the feeble bulwark, cracks appeared here and there, but, luckily, this time the embankment stood the test and did not collapse. Only a few waves reached over its top, and very little water seeped through the weakest, thinnest parts of the outer embankment, barely wetting what was behind. All in all, it had been a lucky night.

Later that morning, I chatted with Nur in front of his house. A matter of pride, and a place of refuge for his large family, it was particularly exposed, right on top of the outer ring embankment. The position of their house signals the volatility of their presence. His life on the embankment was always mired in uncertainty. He had no title to the land he was dwelling on, and depended on allegiance to a *neta*, a political leader, who had died since, and a party, the Communist Party of India (Marxists)—the CPI(M)—whose fortunes were sinking, possibly sliding into oblivion after their defeat in the 2011 elections. But coming here, delving into environmental and political volatility, he had fled the certainty of being in the way of encroaching waters and shifting shorelines.

Earlier that morning, the waters had reached his house. They were close, only a short distance left before the waves entered the two cramped rooms. Fortunately, they were merely lapping at the foundations, wetting the court and encircling the coop. The previous night, Nur too had hardly slept. It was not just the roaring and gurgling of the waters and the shouting of people moving along the embankment that kept him awake. As everyone else in the colony, he had been aware that a spring tide was on the way. The tensions triggered by its onset had worked him up, and the susceptibility of his home and the volatility of his family's existence made him join the crowd of people tending to the outer line of defense. It was a night of danger to his modest house, and also a moment when his existence as a not yet fully legal settler had been at stake. Had the house washed away, his hopes of being integrated into the colony may have dissipated with it.

Soon, the peak was over. The water began to withdraw with the ebbing tide, leaving behind soaked mud, stranded boats, drenched mangroves, and tired but much relieved villagers. The embankment had not collapsed. Battered it was, and a far cry from the elaborate engineering marvels safeguarding coastlines on more affluent shores (see chapter 6), but it had still withstood the waters. Once again, and hopefully not for the last time. The night and the morning after revitalized beliefs about safety. Brackish waters were all around, enclosing the colony in all but the northern direction. However, they could be kept at bay, expelled back into the sea beyond swamp and beach dunes. The colony had kept its promise for a long while now: it proved to be a safe haven from the hardly visible but disastrous encounters with saline waters. Once again, it appeared as a settlement set apart from the perennially flooded, liquifying landscapes Nur and his neighbors had fled. Thus, the auspicious outcome of the night—its uneventful passing without any devastating flooding and withdrawal from the sea—only underwrote a frail sense of security. It fed into residents' anticipations of being safe in the future and buttressed their hopes of remaining unvisited by the destructive forces of flood waters raging in the days to come.

Spring tides, oceanography tells us (Garrison 2011, 232–234), are fortnightly phenomena, when, triggered by the moon, waters rise particularly high. For the delta seafront's fortnightly spring tides, those occurring during the full moon are significantly stronger than the ones occurring during the new moon. I consulted the tidal charts stored in a dusty archive of sorts crammed into a wooden cupboard of Sagar's more than 150-year-old lighthouse complex. The differences between water levels during the two types of spring tides differ only slightly in some months of the rainy season, but markedly in others. Given the sorry state of coastal protection in these parts, so much became clear once again while I was talking to Nur on that day—even a slight difference could make all the difference.

Even though Nur and his neighbors had weathered this night rather well, it underwrote specific horrors of the full moon along eroding banks. The night had brought back memories of other, less lucky times, of spring tides rising with the full moon and breaking through ring embankments, flooding fields, devastating homesteads, and washing away dear lands. Or of slightly more elevated lands being undermined by treacherous currents that would moan loudly immediately before imploding and crashing into the waters waiting underneath.

This experience was not unique to this night. With every full moon, memories resurfaced of disastrous events undoing lives, of being pushed around callously by hostile waters rising and devastating with phases and seasons. Still-fresh memories returned of those long years of living under threat, of being encircled by an encroaching sea pushing inland, eventually turning proud farmers into landless, displaced paupers. It made present those dark times of being at the sea's whims, of being visited by disaster. It invoked experiences that were at once similar and shared yet widely distributed and uneven. Disaster always seems to visit only specific, frequently far-flung shores at the edge of coastal islands, areas known as flood zones. While there is a consistency in this aspect, capricious rhythms and times of intensification quell the potential for a wholly collective experience or blanket theorization. To live through coastal erosion is to be susceptible to the tightening and tensing of the conditions, to its crystallizations into intervals of utmost danger and dramatic losses on the most domestic of levels.

Moreover, full-moon nights lend immediate urgency to otherwise rather invisible processes and may leave whole shoreline stretches bruised. They also generate a sense of shared threat. Spring tides hit all shorelines and all at once, compelling islanders all along the coast to weather them with whatever they have at hand. In the tides' wake, people update on one another and swap stories about how they were faring during particular spring tides, or how protection measures need be vamped up before the incumbent tide draws near.

In rehearsals of past displacements, full-moon nights therefore feature prominently. If this articulates ecological interrelations, it also serves as a dark undercurrent to the festive and auspicious character of full moons in the Bengali calendar. In Bengal, as in many other South Asian societies, full moons structure the year. The sequence and precise dating of most ritual activities and festivities are directly linked to the moon (Michaels 2006, 337). Full moons are times of intensified ritual activities (Freed and Freed 1964; Babb 1975, 125–127), with annual festivals for particular deities being related to a particular full moon. Full moons are understood to be fortunate moments, potential gateways to divinity. The latter conception in particular allows for certain parallels with Hindu notions of space: as much as specific places—the *tirtha*—are treated as overlaps or entanglements of different worlds or orders and hence allow for an increase in the "fruits of worship" (Nicholas 2003), these orders also come closer at specific times.

This approach pervades life in the colony, where, for instance, two specific full-moon periods are climaxes of festivals for the goddesses Manasa and Sitala (see chapter 7). During and after the rains, the festivity is turned on its head. The focus of the residents is then directed toward the embankments, the hazards at sea, and, of course, past suffering triggered by full moons. This double role of auspicious moment and intensive threat emerges poignantly when the full moon marks the end of a monsoon: the Kojagari or Sharad Purnima. The latter is simultaneously the day of the Laksmi Puja and of one of the most fearsome spring tides. Nightly vigils to honor the goddess and tend to the embankments coincide tensely on these nights: they are as much a time of ritual promises, of love or intimacy with deities (Nicholas 2003, 18), as they are moments of danger and loss.

The noises and rush along the embankments call attention to the multisensorial quality of the distributed disaster—one that stands in tension with the notion of its impaired visibility. A word of caution is due: in referring to visibility here, I do not restrict myself to the sensory domain of gaze as such. Unsurprisingly, in an intellectual context favoring access to the world via the eye and associated prostheses (Jay 1993), vision remains crucial in my account. But thundering tides, noisy nightly activities, the smell of spray, and soaked swamps make times of danger a more-than-visual affair. It is this sensory richness that marks full moons as moments not only of risks but of remembering past threats and losses.

But at the bottom of it, that night also underwrote once again that Nur and his family and neighbors were still firmly in the grip of coastal erosion. Yes, they had been left dry for a number of years now and had not had to move home. But with every rainy season, and particularly with every full moon during the rainy season, waters were drawing close, and outcomes were uncertain. The presence of the estuarine waters was palatable in their lives. The disaster lay dormant for now but was never really overcome. Spring tides thus lay bare the temporal rhythm of distributed disasters: waxing and waning, cyclically intervening into lives, threatening and often-enough disrupting routines and eroding land. Islanders are of course acutely aware of coastal erosion's paradoxical patterns, its seasonal and fortnightly rhythms. They speak of lives hampered by living through *bhangon,* the breaking, which visits them in a cyclical fashion. They respond to coastal erosion's temporality by pushing for better protection in the months leading up to the rains, the season of risks, as I will explore in greater detail in chapter 6. But they also account for specific material qualities of shores mediating the onslaughts of spring tides, and, more generally, the texture of coastal erosion. In the legendary flatness of the delta front, the smallest of elevations and compactions chronicle socio-ecological entanglements, bearing consequences for living with encroaching seas (Finan 2009; Colten 2006).

Erosion on Low and High Land

Nur has always lived in what islanders call "low lands" (*niche jomi*), and still does. The land surrounding the artificial embankment he now calls home also belongs to this category. On low land, the landmass's surface sits below the high tide line and frequently also below the mean sea level (see figure 3.1). Geomorphologically, such land is the youngest. Until settlement, tidal waters would enter freely and leave sediments behind, thereby continuously but slowly accreting land. Diking terminated this process. In keeping waters out, embankments promised to enable agricultural futures by allowing for the sweetening of the land. Simultaneously, they also endanger dry land-dependent forms of life. Soil dynamics at the delta front makes this clear. Once tides are kept out, what ceases is not only the deposition of sediments but also the gradual raising of the surface and a regular replenishment of its fertility through the sediments left. The lands are now dry for most of the year, and not soaked twice daily with tides washing over them. Drying, in turn, compacts sediments, both at the surface and below. In due course, the soil layers compact in the upper strata, which adds to the "normal" process of compaction of deeper Holocene layers (Nandy and Bandyopadhyay 2011). In the process, the surface of embanked terrain subsides—considerably faster under these anthropogenic conditions than would be the case otherwise.

Through this double process of forestalling the growth of the island and fueling its subsidence, formerly well-watered intertidal lands turn into saucerlike plots enclosed on all sides by embankments (Jalais

Fig 3.1. Erosion on low-land shores, Botkhali Island, Sagar. Photo by author.

2010b). If sea level rise is added to the sinking of compacting soils, relative water levels are rising even more dramatically (Hazra 2012) and floods are becoming increasingly more destructive as they gush with greater force into subsiding lands and remain trapped here for longer periods of time.

On low-lying coasts, coastal erosion coheres around events and processes. Its processual, distributed nature is mirrored in the polyvalence of the concept *bhangon,* erosion, literally "the breaking." Let me unpack them one by one. To Nur and others, *bhangon* refers to the undoing of outer embankments by relentless waters. Waves attack these mounds, whipped up, they say, by tides and winds. Currents gnaw on them. Most of this activity is bare to the eye, occurring on the surface, in contrast to the largely invisible action along high lands, as I show in the next section. Weakened embankments eventually collapse, most often during periods of heightened stress. Failed embankments serve as entry points for tidal waters that swamp their immediate hinterland through small-scale, contained floods. Given the shape of the land, water enters easily, but outgoing tides rarely fully leave, with devastating consequences for agricultural prospects. *Bhangon* captures the combination of collapse and chronic swamping by floods, occurring well below the level of headline-grabbing floods.

Such flooding following embankment collapse is an integral aspect of the estuary's advances on low land. It signals that even in a climate that reappraises fluidity and amphibious life (Wakefield 2019; Krause 2017), floods continue to wreak havoc (Camargo and Cortesi 2019). In doing so, they also add to the list of specific flood types. However, the anthropological and geographical accounts demonstrate that, to farmers in the Bengal delta, floods are not generally problematic. In fact, in fresh water–dominated parts of the delta, floods used to be very welcome for replenishing fields with fresh sediments and ponds and paddy fields with fish. In short, floods enhanced fertility. People here would distinguish between good, nurturing floods (*barsa*) rejuvenating the land and destructive floods (*bonya*), which spell trouble by reaching too high, staying too long, or occurring at the wrong time of year (Rasid and Paul 1987; Schmuck-Widmann 2001).

Closer to the sea, in the Bangladeshi Sundarbans, the terminology is slightly different. People here differentiate between three types of floods. They know *borsha,* the floods brought by annual monsoon rains. They know *bonna,* floods following surges and embankment collapse. And they know *jalabaddho,* waterlogging when monsoon waters are blocked from draining into rivers. The last one is, as Camelia Dewan (2021) has meticulously shown, a recent phenomenon that became rampant as a consequence of thorough embanking of the delta. Blocked from getting spilled across the land during a flood, the sediments stay within the rivers. The roughly one billion tons of sediments annually washed into, and through, the delta (Islam et al. 1999) clog whole rivers, raising beds and blocking channels. Which, in turn, traps excess water in instances of waterlogging.

On Sagar, within the immediate reach of the tides, people distinguish between still other types of floods. My interlocutors differentiate between rain-fed floods (*akash bonya*) and river floods (*nadi bonya*). They consider rain-fed floods neither destructive nor beneficial. Excess rain water does not fertilize the soil but can here, right on the sea, easily be flushed out during ebb tides. River floods, on the other hand, are generally haphazard and further subdivided into two types: floods after surges and normalized inundations of particular strips of land behind collapsed embankments. Considering the latter type as floodings not only calls attention to the little-researched topic of small-scale repeat inundations at the marine edge, it also captures processual remodeling of terrain and the swamping of agrarian lifestyles. Thus conceived, floodings complement the process of seepage (Cons 2017), where waters ooze through embankments without collapsing them. Indeed, both feed into one another, as puddles of water seeping through might signal imminent embankment collapse while a dotting of puddles indicates floodlands during low tide. Both border the nonevent while embodying potentially calamitous changes to people under conditions of neglect.

Bhangon also refers to the uneventful undoing of landscapes between the moment of embankment failure and the moment of its reconstruction. It refers to a time and space where brackish waters enter freely with the tides, washing out crops or homes, reducing the quality of the land the longer they enter, turning village life for those who stay into an ultimately amphibious one.

Finally, *bhangon* on low-lying land refers to the ultimate disappearance of abandoned land once ring embankments have been shifted inland. I explore the specific social nature of *bhangon*—that is, the moment of abandonment instantiated by moving ring embankments inland—in chapter 6. Here I want to highlight that erosion is subject to the temporalities of seasons and tides on the one hand, and of temporalities of repair and abandonment of protection devices on the other. It results from an intertwining of environmental and political processes.

Most shores on and around Sagar belong to the low-land category. Some, however, reach above the high tide line, notably on Ghoramara. Islanders refer to such banks as "high land," *uccho jomi*. Smaller embankments line high-land shores, in place to safeguard the land from spring tides and rough seas during winds but not from tides. Coastal erosion on high land proceeds differently than on low lands. Here estuarine water persistently claws away at exposed banks (see figure 3.2), until smaller or bigger chunks break away for good. Islanders, however, insist that on these shores the most consequential erosions occur surreptitiously. Below the surface, currents appear to be of greater velocity, and loamy mud gives way to deeper layers of sand, which erodes quicker. Thus, land at the edge may cave in anytime, and erosion here is more chronic and less mediated by embankment collapse or flooding than on low-land shores.

On high land, my interlocutors were more concerned with debating when plots were likely to go instead of discussing when the embankment would collapse, as they would on low lands. This strategy underlines a critical difference in the anatomy of erosions in respective types of shores. On low land, erosions are, as noted, frequently precisely timed. They manifest in often exactly foretold and pinpointed embankment collapses, in small-scale flooding events, and following land abandonment after withdrawal of ring embankments. Coastal erosion often proceeds

Fig 3.2. Erosion on high-land shores, Ghoramara Island. Photo by author.

there in well-timed, rhythmic steps. On high land, however, land is under-mined and silently washed away all the time, and the timing of erosion is more elusive. But even if chunks may break away anytime, there are rhythmic intensifications, and erosion is seasonally inflected. Outflowing tides, particularly during their peaks, hit slopes the strongest, loosening sediments more rapidly. Seasonal spikes during the rainy season or spring tides also increase the velocity of erosion on high land.

Furthermore, erosions here may be condensed into short events. Many people referred to instances when larger chunks of land collapsed suddenly. Undermined by subsurface currents, sometimes whole plots would implode and be sucked away in an instant. Horrifyingly swift, such cataclysms were still announced by visible cracks in the land and loud noises, and they occurred in areas of risk. Most instantaneous shifts occurred in areas considered acutely eroding, and thus within the con-tours of what I call deafening certainty.

The Tree and the Children

In the early days of my research, when I was figuring out the pre-cise mechanisms of coastal erosion along low and high lands, my inter-locutors would frequently invoke one particular incident. Their recounting of it deftly illustrated the violent horrors of land being undermined and breaking away in large chunks along high shores. I call this incident the story of the tree and the children. Throughout my fieldwork, islanders returned to this story. It seems to embody the coming together of differ-ent processes into an event of dramatic proportions, where something hardly visible was primed for storytelling and intersubjective coping. In fact, this story is one of the moments through which dispersed people and distributed disasters are drawn together and aligned in a single nar-rative and, thus, rendered a collective, witnessing drama and death.

Many knew of the tree and children by hearsay, and few had wit-nessed the proceedings firsthand, but eventually I did speak to one of the survivors, Sheikh Mumtaj. The basic structure of the tale remained unchanged across its many iterations. It kicks off with water circling in on a large landmark Bodhi tree standing on high land on what was then the edge of Lohachara—the subtext being that, meter by meter, the water had gobbled up surrounding terrain and the trunk had ultimately become part of the shoreline. The waves were then gnawing on the roots and the trunk, and the tree's collapse seemed imminent. Close by, a group of islanders had gathered to watch. Two boys were playing very close or even underneath the tree. Sheikh Mumtaj, one of the two boys, told me that their mother had forbidden them to go near the tree. But the brothers were curious as to what this was all about and what would happen to the tree. Secretly, they went. And as they got very close, the towering giant gave in and crashed into the waters. Most people were watching safely from a distance, but the boys were right there as it happened. The tree took them with it. With nothing to hold onto and no refuge, they were sucked into the vortex. Only Mumtaj managed to reach the surface and escape the whirls and currents. His brother was lost. He disappeared into the deep waters, his body never found.

It was a shock to the surviving brother and his family. The waters left them bereft of a loved family member. It was a shock to the islanders who had helplessly witnessed the tragedy unfold. In conveying existential loss, the story, I suggest, crystallizes events otherwise hard to convey. The swaying tree, taking the kid with it as it collapsed into a watery nothingness, is a climactic moment ripe with loss, powerlessness, and bare survival. It had disastrous dimensions and encapsulated the shared fears of islanders. I contend that these dimensions help to account for the story's popularity.

That being said, details changed across iterations. And some of the details matter. Consider, for instance, the precise timing of the event. A number of my interlocutors would not identify the incident's date or even the season. Others, however, did. Sheihk Mumtaj, for instance, insisted that waters were particularly ferocious because it was Astami. But as the Astami (Sanskrit: $aṣṭāmī$) marks one of the key days of the Durga Puja[1]—a major Hindu festivity celebrated lavishly across Bengal—death and narrow escape came to be associated with a day of great significance and ritual auspiciousness to at least the Bengali mainstream (Rodrigues 2003, 71–247). Even if unmoored from festive full moons and untouched by the troublesome Ganga (see chapter 7), this timing once again underwrites the dark undertone Hindu festivities carry on these shores.

---

1. Technically, the term $aṣṭami$ refers also to the eighth day in every fortnight of the Hindu lunisolar calendar. According to the latter, every lunar month is divided into two halves beginning with the new moon and the full moon, respectively. Not only fortnightly intervals but also every day within the fortnight of a certain month is assigned with particular meanings, rules, and regulations (see for more details Freed and Freed 1964; Wadley 1983). That being said, in colloquial conversations in Bengal, the notion $aṣṭamī$ is used to index one of the peak days of the Durga Puja, which falls on the $aṣṭamī$ of the solilunar month of $aśbin$ (September/October) and is often also called $Mahāṣṭamī$ among devotees.

Yet, on several occasions, I was told that the children had simply been playing on the shore when the tree capsized. These iterations allude not so much to an event foretold—anticipated by the onlookers gathering curiously and by the mother forbidding her kids to go near—as to one of chance. My interlocutors would emphasize the element of surprise, the roaring sounds, and the very transformation of the landscape in an instant, to the effect of evoking swiftness along with a sense of victimization and powerlessness.

Unfailingly, however, the story would begin with a reference to that tree. Not a tree, but *that* tree. It must have been an impressive specimen, one of the few Bodhi trees that stood on the young, now sunken island of Lohachara. The tree today serves as a landmark on mental maps, orienting and arranging bygone days, homes on long-gone islands (see, for instance, the opening scene of Saha 2009). Some islanders can even place their former houses in relation to it. It is worthwhile to reflect on the role of the tree within the narrative. Bodhi trees rank among the most widely valued trees in India (Haberman 2013). Alongside ritual dimensions, they are seen to embody community and are imbued with a sense of belonging, featuring as central places across India, with community afforded quite literally in their shade. For colonial observers of India, the entanglement of villages in the air roots of giant Bodhi seemed to encapsulate what they perceived as the grotesqueness of the subcontinent (Rycroft 2006). In the delta, trees are among the sturdiest and longest-lasting materialities available. This is why they have been climbed during cyclones for centuries,[2] a pattern evoked effectively in the climax of Amitav Ghosh's (2004) novel on people and tigers in the Sundarbans. Trees represent all that waterscapes are not: rooted, fixed, durable. Meaning is attached to that Bodhi tree, a now-absent entity that has become an affectively loaded landmark for navigating sunken terrains. Against this background, it appears safe to read the story of the tree and the drowning children as one that speaks to the perceived threat to village communities, and to fears of seeing rather durable materialities and socialities being pulverized.

Death lends tragic depth to the story. It is tempting to bundle up islanders as a whole alongside the children, to portray them as innocent and vulnerable, as Sheikh Mumtaj had. In fact, media accounts regularly

---

2. See file "Cyclone, Storm etc. 1956–1979," Marine Department, Marine Archives, Kolkata.

take on this approach when they explicitly seek to portray displaced islanders' children or grandchildren in order to convey a sense of powerlessness and victimization (Harms 2018). However, the narrative binds three distinct groups of actors together in a moment of foreboding and dramatic eruption. The constellation of tree, children, and spectators is emblematic, I argue, for the way islanders living through coastal erosion see themselves: as silent onlookers who know very well what befalls them; who see what is about to happen and when, yet cannot muster the means necessary to protect what is most precious to them; who are helplessly staring as coastal erosion threatens what is most precious; and who are suffering a form of environmental violence yet—due to their environmental knowledge and temporal envisioning—are almost always in a position to move bodies and immovable property out of harm's way. In other words, safely removed, hardly seeing their corporeal integrity threatened, the onlookers are still powerless.

It is here that the story's power unfolds, which may explain its popularity among islanders. It is a telling rendition of a collective envisioning itself as reduced to a witness of the literal dismantling of the landscape of daily life, a dismantling that turns deadly only to the weakest members. The constellation is also emblematic of the temporal pattern of coastal erosion, calling attention to spikes and deflections amid drawn-out stretches of crisis. In this view, the story achieves a rare feat. It renders erosion a frightening and violent event, condensing tragedy and freeze-framing drawn-out processes. It calls attention to collective aspects of experiencing erosions—by implying a group of onlookers as well as by drawing narrators and listeners together into the shared space of the story. This reading is strengthened by the victims' emanating innocence and limited power.

Moreover, it resonates again with the above-mentioned metaphor of the "shipwreck with spectators" that philosopher Hans Blumenberg (1997) traces across the history of European thought. Blumenberg argues that, rather than emphasizing despair at the inability to help, the metaphor of people watching a shipwreck from the shore actually articulates content on the side of the witnesses for being themselves safe and sound. Contrary to Blumenberg's account, the onlookers in the story of the tree and children appear safe and sound only for the moment. What might seem like restrained relief about being on firm ground, out of reach of calamity, is caveated by the knowledge that the merciless waters will sooner or later encroach further.

An Encroaching Shore

Much like in the story of the tree and children, recollections of the experience of coastal erosion almost always begin with the advent of brackish waters at one's property, as if appeared out of the blue. But as I live among islanders, probing the past, witnessing the coming and going of freshly displaced people, as I relearn the landscape, the deep history of erosion (and accretion) on the delta front dawns on me. I have never met a person who could qualify as the first victim of erosion on these islands. Maps and histories help. They tell of the ongoing nature of the blows sustained by the estuary and underline that at each moment in recent history the land on the shoreline has been a frontier, dividing a not-yet-affected area from already-abandoned or swallowed lands. They tell of villages housing hundreds swept into the sea, of the march of the waves, of currents cascading through once-lively and fertile lands. On Ghoramara, my interlocutors witnessed the demise of entire settlements, and, finally, the total submergence of the islet of Lohachara just off its shore. Only villagers' narratives tell of their existence, alongside local history books, maps, and a few scholarly works. Coastal erosion is not a moment of destruction, nor a scenario where people return to nothing (Read 1996), but a dawning of disruption and displacement, a process by which the watery expanse tightens its grip. The shoreline turns out to be a moving frontier, pushing inland and slowly exploding into the lives of individuals, families, and villages.

Following my account of full-moon spring tides, of local taxonomies of shorelines, and a story of death, I will jump back in time only to work myself forward chronologically. I begin by detailing processual dynamics, unpacking erosion as condition before turning to what I like to think of as moments of crystallization. This engages with what I understand to be the standard narrative of environmental displacement on contemporary Sagar and adjoining islands, probing its contours, silences, loopholes, and openings. I turn to life histories marred by coastal erosion, unraveling narratives and silences, to explore how the distributed disaster is situated and selectively remembered. Once again, certain times and certain episodes stand out while other, more demeaning or chronic modes of living through coastal erosion seem to fade. This poses a puzzle. Why is it that islanders remember selectively and often remain rather silent about the long times of going through what to them counts as the true disaster?

Writing on living with erosions on the sandbars far removed from the shore, the *chars*, in central Bangladesh, Naveeda Khan engages a similar problem. Here, too, people consider the dismantling of homes and lives and relations by erosions to be calamities, as "death, as an excision from a place, time, and milieu" (Khan 2022, 68). Yet, what it was to go through these troubles, how it felt, and what happened recedes, she notes, into the background of conversations. This is in marked contrast to elaborate conversations on what it was like to go through cyclones affecting whole populations and becoming tied up with events of national history.

Khan reads the silence of *char* dwellers on erosion as an "obscurity to themselves": "in the throes of erosion [they] were not self-conscious, much less historical, subjects. Rather, chaura's [*char* dwellers'] obscurity to themselves indicated their state of receptivity and thereby their transmission of the river's activity." Taking German philosopher Friedrich Schelling's interest to theorize how nature works through humans to the Bengal delta, she thinks of such "self-opacity as receptivity toward the landscape" (Khan 2022, 61).

On Sagar, I encountered a somewhat similar "self-opacity" operating in retrospect as people attempted to come to terms with the onslaughts of estuarine waters. Here, too, silences set in that stand in contrast to the dramatic effects *bhangon* had on lives and fortunes. I distinguish two conflated types of silences. One shrouds the estuary's gradual encroachments on one's life eclipsed, as it were, by the drama of seeing one's home attacked. The other relegates into the background the long years when people saw others lose their homes to the water, long before it was their own turn. Complementing Khan's take, I read these silences as signaling the humiliations and the sense of futility emerging from failed attempts to avoid displacement by keeping the waters at bay; and as going through these on individualizing terms—one house after the other—not knowing what it is that will befall them, instead of a larger collective sharing the burden of experiencing suffering and making sense of it.

When pressed, no one denied that erosion had been happening before and elsewhere. But it became a condition shaping my interlocutors' everyday lives precisely as brackish surface waters reached the land. In retrospect, it was a moment of shock, something that set in suddenly, as Ashini Pal, who was among the first to settle in Colony Para, explains.

Alongside conveying the individual effects of coastal erosion, the silencing of the shore's slow advance onto lands and into lives highlights a degree of denial of one's ever ending up at the frontlines and right in the path of the estuary pushing inland. Many villagers stick to their land right up until the very last moment, hoping that the destruction of their lands will cease eventually or comprehensive solution will be put into place by state agencies. As they stay on, the waters close in on them. In short intervals, the distance between the shore and the houses diminishes. In some places, the waters draw closer with every rainy season. In many others, they do so only when embankments are shifted every few years. In either case, with the waters drawing in, outlooks diminish. For one, the economic value of plots declines. After all, who would pay decent amounts for land earmarked for disappearance? And who would enter lease agreements in any but the most unfavorable conditions with such a piece of land? Similarly, the agricultural productivity of land is at risk as the shore increasingly draws near. Long before brackish waters enter fields as floods, or break high lands chunk by chunk, saline water incursion takes its toll on the ground. And so it did with Nur. He felt ensnared, with nowhere to go, forced to hold on to the little security he had. Alongside others, he was trapped.

But when the grounds between river and field had literally evaporated, and the shore began to encroach directly, environmental relations turned critical. People found themselves in the grip of erosion. The shore makes an appearance on the outer limit of people's lands, entering fields for the first time, or, on high land, beginning to break away chunks. In most tales, this marks the moment when erosion began, and is a threshold against which to gauge meaningful and manageable drawn-out processes, highlighting the individual effects of the distributed disaster.

In narrations, coastal erosion appears to be unbound from here on. Echoing what everyone else said, Nur distinguishes between more and less risky times, between times of affliction by greedy waters and times of comparably comfortable quiet. Tension is contrasted with rest, yet coastal erosion is never fully absent. It is a force to reckon with, a reality to be handled by keeping a distance from the waters, and by making sure that the distance is held intact. A good life, then, is also a function of living as far as possible from the sea. The proximity of the waters is what keeps people awake at night and drives them out onto the embankments.

The complex temporal and spatial pattern of erosion reappears at the domestic level all over again. Landholdings in the Bengal delta are known for being fragmented and dispersed (Banerjee 1998), with Sagar being no exception. Many of my interlocutors used to own various plots, spread out across villages. Both a strategy for adapting to haphazard rains or river floods (Schendel and Faraizi 1984) and an outcome of inheritance laws, fragmentation means that landholders do not go through the process of coastal erosion all at once. Sometimes, years of quiet, out of erosion's direct reach, are interrupted by episodes of acute submergence. One plot emerges as threatened and turns unproductive before being subsumed by the river or becoming buried under the next ring embankment. But with the house moved onto interior plots and with cultivation steadily in progress on other plots, life returns almost to normal—only to be affected again years later. In such circumstances, impoverishment is gradual and distributed over years and decades.

In either case, once the waters have made an entry onto villagers' fields, coastal erosion enfolds affected families as a condition. Across all localities where I have done fieldwork, accretion has never outweighed erosion. The trend—the movement of the water—has been firmly directed inland, the pulverization of lands inevitable. As displaced islanders recount the times that ensued, gloomy tones are tangible. These are narrations of sorrow, uncertainty, and bitter struggles against an overpowering sea. Often enough, the narratives are laments shot through with bursts of attributing blame. Yet, overall, these were years clouded in silence, years of muddling through and working the lands against all odds.

At present, islanders caught up by the edge continue to experiment with putting their doomed plots to use. On high land, this means planting rice in the parts of threatened plots far removed from the water, to ensure at least some harvest. Costly seeds and fertilizers are omitted, so no considerable investments are wasted once the plots disappear. On low land, this may mean resorting to saltwater aquaculture, however, most people only increase their dependence on other sources of income, such as fishing or circular labor migration. Being in the midst of erosion entails increased dependency on extra-agricultural incomes. Such a dependency certainly is the mark of India's extended rural crises (Shah 2011; S. Bose 2007). But erosion adds to the injury. Looking back at failing harvests, diminishing outlooks, and repeated floods conveys a sense of individualized helplessness. These are uncomfortable, demeaning pasts, extended periods of nonevents, frequently pushed into oblivion.

Much like spring tides and seasons, certain climactic moments stand out within these gradual shifts. They juxtapose the relative mundaneness of normalized landscape transformation in individual trajectories and life histories. Being in the midst of erosion is therefore not only a condition to bear. It is interspersed with further personal moments of heightened threats, times when slow erosion comes to be dramatically visible, when the encroachment of the waters forces islanders to act decisively. If coastal erosion was to be visualized in the form of a graph, the moments I am turning to now would emerge as rapid spikes. Such moments feature prominently in recollections of pasts or presents troubled by encroaching waters. This prominence may answer for difficulties in speaking about,

relating to, and remembering, hardly visible transformations. It may articulate some of the pitfalls of rhetoric and the prioritization of drama in
ordinary language and especially in communicating to external audiences,
including the odd anthropologist. But most importantly, I suggest, it rests
on a sense of futility and powerlessness. Slow deterioration hardly leaves
room for agency. Its uneventful nature brings a series of barely visible
deteriorations, enduring quasi-events, and anxious waiting (Auyero 2012;
Povinelli 2011)—whereas cataclysmic events, in their drama and plight,
lead to stories of demise as well as survival and success. It is for this reason that erosion as condition proves once again to be more challenging to
describe—for my interlocutors at the forefront of this watery advance and
for me as I write this book. To focus on particular thresholds provides
some grounding.

Letting Go of the House

Within recollections of life under coastal erosion, abandoning one's
home comes to be the defining moment of despair. Displacement lurks;
thereafter, islanders turn into landless paupers, or so they see themselves,
now dependent on the whims of overpowering waters, and whatever
petty gifts by an elusive state bureaucracy come their way.

These are moments of flooding and flight, frequently involving
planned abandonment and often also rushed retreat in acute moments
of danger. Nur, for instance, returned to this experience many times. His
family had lived in close proximity to the embankment for a while, where
they had weathered, he insisted, many floods. They were used to it. "This
is how we live in this flood-prone zone [bonya elaka]," he said, "in this
salty land [nona jayga]" on the edge of the island, resorting to an unusu-
ally somber tone. Like most people, he knows how to make it through
floods. They would store important papers in several layers of plastic bags
hung from the ceiling and valuable items in watertight containers during
seasons of risk. They would stay put and keep to the bed in flooded homes.

The eventual demise of their houses turned out to be no surprise to
most of my interlocutors. Cracks appeared and the floor turned rough
due to the salt. Prospects turned ever bleaker by spring tides, the roaring
waters constantly drawing nearer—so much I could reconstruct across
extended conversations. Eventually, most families decided in good time
that it was upon them to move, that the repairs were not worth the effort
in the face of an ever-nearing sea. If a new place had been found, belongings were shifted well in time and essential movable parts of the house,

such as tiles, poles, or doorframes, were frequently taken along. Over the years that followed, such parts would serve as frames for new houses or makeshift huts.

Others, who either stubbornly held on to their original home or who lacked the connections or financial means to find a suitable place elsewhere, stayed to the very last moment and were eventually ousted by the floodwaters themselves. Nur's family belong to this category, literally flushed out of their home by unrepenting waters.

Nur's account of evacuation is rather atypical. His home collapsed swiftly, he vividly explained, and rather out of the blue, during one of the spring tides. He habitually insisted on the slowness and encroaching nature of *bhangon,* narrating how long periods of his life had taken place, and still are, in the midst of it. Yet, he was startled himself by how this slowness could quickly evaporate, replaced by an onslaught demanding swift action. One key moment was the day his family lost their house. "It did not even need fifteen minutes," he said.

> Eighteen *bigha* were lost at once. From that day on, it all belonged to the river . . . It was not during a storm, but during the springtime of the month of *bhadra.* We had this hut, and we stacked rice and other things on its roof. In the daytime, a little water flooded, so I knew that at night it would be all sinking in water . . . But it was different. It came very quick. The embankment broke and salty water entered all around. The house broke, and the things we had stored on the roof were lost. We lost voting cards and foods. Later I recovered the rice [in the water], but it had become too salty to eat.

"But listen," he concluded, "nobody could know how quick this was about to happen. I had taken a good look at the embankment and seen that it was good. Something would be disappearing [namely, eroding] at night, I realized. That was the situation. But then it came so quickly!"

Much of Nur's account cohered around this incident. The swiftness of the waters, his family's scrambling to hold on to what could be salvaged, the abandonment of their home. Even though they had been within the fold of coastal erosion for a while now, it was a moment when it struck rapidly.

In Nur's account, the spring tide and the collapse of their home fell into one. This holds true for many accounts. To most of my interlocutors, however, collapse did not come as a surprise. In fact, the collapse remained a critical element, one that was partly expected but nonetheless panic inducing. The loss of the house both was and was not a surprise. Although struck by the swiftness of the waters, they knew for a while that it was coming for them.

Among my interlocuters, the demise of their original homes set forth a series of retreats rather than escapes. In most cases, they were decided upon in advance. This stands in contrast to media representations, where villagers are portrayed fleeing in the dead of night or framed in the language of the refugee, a depiction wrought with victimization. Even if surprise at the swiftness and timing of collapse were absent from most accounts, the moment of collapse remained decisive for what it spelled out for affected families. It still accentuates an instance of deeply felt loss, a moment of disastrous proportions. This event—always personal yet situated within a shared trajectory—turned farmers and landholders into the displaced. It is a moment of negation, heralding times of deepened misery and aggravating struggles to make ends meet, more than an event triggering flight, panic, or fear for survival. This moment also accentuates a deep rift between the homes abandoned and the many places of refuge sought thereafter.

As I listen to my friends and interlocutors narrate their pasts, the hopeless sensation of having nowhere else to go and feeling trapped was met by an underlying wish to stay close. They want to stay in the vicinity, my interlocutors tell me time and again, close to their neighbors, close to people they know, to people who can relate to their predicament, who know them as more than simply landless misers, who know them for what they were until very recently: farmers and landholders. They bet on a sense of solidarity or, at least, of being tolerated. Equally important, they stay put in order to qualify as beneficiaries for state rehabilitation schemes, to prove their eligibility by virtue of inhabiting this wasted landscape. Not merely "trapped populations" (see Black and Collyer 2014), they also chose to stay put as a form of betting on future benefits and, as I show in chapter 5, for attachment to place founded on shared deeper histories. But as they do not move far, the shore draws in on them again. The water catches up with them, people pack up, move on to a spot nearby, and rebuild houses. Until the waters draw in yet again. And again.

Most of my interlocutors have moved house like this five times or more. As the waves draw close on them, they dismantle their homes only to rebuild a shadow version of them on another line close by, until they managed to secure land elsewhere through either governance provisioning, squatting, or scraping together funds to buy a plot elsewhere. Many, therefore, engage in what I like to think of as micromigrations: shifting with the receding shore, rebuilding homes and lives in the vicinity, in a modality of movement that is in itself rhythmic, and sits between staying anchored in place and moving out for good.

There is a deep-lying awareness of being under imminent threat, of once more having to hastily pack up belongings and rescue all that could be carried along. They are perpetually driven by footloose shorelines, with immobile assets liquefying and neighborhoods washed away. Shifting into the interior of the island, they made use of a sense of shared predicament while seeking refuge, a sense of being in the midst of it together. Yet solidarity hardly stretches far, with barely anyone taking notice or extending care beyond the immediate neighborhood.

Listening to islanders telling their life histories (*jiboner itihas*) reveals a fairly uniform narrative. It is a narrative shaped by coastal erosion and neglect. The preceding section has highlighted three defining features of this narrative: the arrival of erosion as a shock, the first loss of home, and recurrent relocations thereafter. All of these tropes are bound up with specific silences. Indeed, these common features become what they are only through the workings of silence. They virtually erase either the slow encroachment of waters or the rather demeaning periods of people making do as barely tolerated squatters on public land. Taken together, this narrative heavily emphasizes the individualizing effects of coastal erosion.

As patterned narrations, the recollections point to the socially mediated character of remembering as well as the critical role silence plays here. After all, as novelist Indra Sinha (2008, 47) noted, "silence is what makes sound into song." More specifically, these narrations articulate in themselves ways of coming to terms with unbound processes distributed unevenly across time and space. Stopping short of providing a way of attributing meaning, such a narrative still articulates and inscribes an experience of a shared predicament to people or small groups affected under different conditions and in different times by something whose very nature is not yet resolved. I engage in chapter 7 with how islanders attempt to explain what it actually is that befalls them. Explanations range greatly, from it being the river or the sea, to ambiguous depictions of a divinity or even the marine trade. Given these profound uncertainties, the narrative frame provides a road map through the literally swampy terrain of the distributed disaster.

In its specific content, this uniform narrative also has moral connotations. More than betraying a sense of being innocent or simply victimized, it coheres around moments of dramatic losses, silencing perhaps even more humiliating years of retreating with the morphing coast while waiting for and dreaming of state assistance. Silences here are not only the outcome of the disappearance of most material markers with which memories otherwise entangle so well (see, for instance, Revet 2011; Ullberg 2010) in muddy waters. If memories are means of rendering present what is long absent, it may come as little surprise that socially mediated recollections cohere around not only specific times but also specific things, such as one's original house and fields, overshadowing roadside shacks many would seek shelter in after displacement. The silence on years spent squatting contrasts again with detailed recollections of resettlement and the making of futures in new colonies and homes, which I turn to in chapters 4 and 5. The silence on an in-between phase, when people were sleeping in roadside shacks and waiting for resettlement, appears, then, as a moment of crafting memories so as to fade out, more implicitly than explicitly, I would suggest, humiliating pasts far removed from local conceptions of the good life and rural prosperity.

A Collective in Time

So far, I have turned to islanders living through multiple displacements by an ever-encroaching estuary. Spectators witness landmarks collapse, a boy drowning, and people struggling, or moving along, with shifting shorelines. These accounts introduce zones of acute coastal erosion as a moving frontier separating murky brackish waters from the domain of the still solid. As the waves and currents hollow out and break high lands, as they wash away low banks elsewhere, the outward sea-facing perimeters are zones where islanders see parts of their land disappearing year after year. A plot on one of Ghoramara's fastest-shrinking edges might disappear within one season. A similar-sized plot at another, less severely affected bank will be diminished only after several years or decades. A family whose landholdings are disaggregated across the island will be in the grip of *bhangon* for decades, delaying the final moment of displacement. In each case, erosion is an individualized and individualizing experience. At any given stretch of the shoreline and at any point in time, only a small set of scattered individual homes are on the line. Consequentially, coastal erosion always displaces only individuals and small family units at any given moment. To be sure, small numbers of families might take the decision to move out of the reaches of the water at the same time, and they may even see all their lands disappearing in the same season. Yet, there hardly is, I suggest, a moment when coastal erosion hits a large group at once. On the contrary, this disaster proceeds relentlessly and in a disaggregated fashion.

Such a scenario belies the conceptual framework of disaster studies all over again as it involves rolling sets of affected individuals, always below the threshold of what could be called a population. At the same time, the language of crisis, chronicity, or everydayness equally fails to account for living through these circumstances. Not only is this disaster spread out across a critical everyday shaken by condensed cataclysmic outbursts, it is also distributed among groups of victims and witnesses. In the absence of one event or a process unfolding in time and enfolding populations all at once, it seems difficult to locate larger groups of the affected. Yet, over time, considerably large populations have been affected. Numbers vary widely. On Ghoramara and Lohachara alone, between six thousand and seven thousand people have reportedly been displaced (Narayanan 2015; Samling, Ghosh, and Hazra 2015), with all other islands in the delta in similarly rough waters unaccounted for.

Like Nur, most of my interlocutors never moved out for good from the shrinking islands or even from their coastal fringes. Regardless of whether flights were planned well in advance or rushed, life was to be salvaged by moving out of the sluggish and overtly salty to firmer grounds. As many others, Nur shifted to one of the interior embankments nearby, doubling as a brick road, calling this "line" home for the next three years. I did not have the chance to visit either the site of the house nor the road as such before they were washed away. But the way in which Nur and his wife spoke of this time in their lives matches well with how other islanders continue living in similar circumstances on these eroding edges. And it is here—in what I understand to be an itinerant slum moving with the shoreline—that the individual dimensions of the distributed disaster converge into a collective predicament. In conjunction with resettlement colonies, such ephemeral settlements allude to a larger collective emerging in time—one that continues to convolute the inner perimeters of eroding zones. People unable and unwilling to leave stick to these frontiers between not-yet-washed-away ruins and still-comfortable adjacent villages. Shelters are built quickly using bricks, wood, tarpaulin, tiles, and bamboo brought from abandoned homes (see figure 3.3).

While they spill onto the brick road, for obvious reasons, blocking it is not an option. Homes are narrow and tiny, lining the road like pearls on a necklace; they are crammed and lacking amenities such as toilets. Living conditions are described as dismal. Informal settlements characterized

Fig 3.3. Freshly displaced islanders living by the side of the road, Sagar Island.
Photo by author.

by a high degree of instability and insecurity, these housings qualify as
rural slums. I use the term "slum" deliberately here to convey the sense
of misery, abject poverty, and unhealthiness evoked by my interlocutors.
Women tend to bear the brunt of this situation. When one lives without
amenities such as toilets, chores are tougher, with frequent longer walks
to drinking water sources. In fact, many of my friends now settled in
proper houses were hesitant to talk about these pasts, burying these
times of helplessness and victimization under more cheerful tales of
resettlement and rehabilitation.

Moreover, these houses appear as footloose settlements. Not only
are they astonishingly shapeshifting, as informal settlements generally are
(V. Das 2011; Ghertner 2015), settlements themselves are mobile. They
move along with the coast, retreating before being drenched by encroaching waters. Unsurprisingly, the movements of itinerant rural slums are yet
again individualized while following similar temporal patterns. Seasonal
risks send ripples along the coast's outer perimeter, making those closest
to the shore pack up and move. Like others, Nur referred to the full-
moon spring tides as times of returning tensions and disrepair during the
years of living on the line, thus underlining the cyclical moment of *bhangon*, erosion, and its freshening of troublesome pasts over the years.

For many of my interlocutors, this was a moment when another kind of cyclicity set in. The wait for the waters to arrive turned into waiting for state assistance, before seeing the waters encroach all over again. In this moment of waiting it out, I suggest, polities emerge. Here trapped populations become palpable, as some people are unable to move out from zones where coastal erosions become acute. Such experiences may appear fairly typical for everyday life on the margins, but to Nur and many others, they came to be framed as a consequence of *bhangon* and government abandonment. In other words, itinerant slums provide grounds for collective experiences to emerge out of individualizing coastal erosions and turn them into political forms of making claims, a key facet of the politics of disasters.

The contours of these polities emerge in different moments. On a day-to-day level, they come up in interactions between hut dwellers. They try to enlist each other's help in order to combat acute shortages of money or foodstuffs, referring to a shared predicament of displacement and impoverishment. Similarly, recently displaced islanders refer to collectively shared experiences—undone lands and broken homes—to distinguish themselves from surrounding villagers. Being allowed to settle on embankments and village roads draws on what Foucault (2012, 84–86) calls "tolerated illegality." Not simply being bestowed, tolerated illegality requires constant crafting. On eroding shores, it takes the form of exclusion of other impoverished ones from the polity, both referring to and thus performing a collective in time and space. Environmentally displaced islanders distinguish themselves sharply from "ordinary" landless people living on the streets of rural and urban India. Whereas those other people ended up on the streets—I learn in conversations—due to bloating family sizes or strife, erosion-affected islanders frequently emphasize their innocence, positioning themselves as worthy of being allowed to stay on. Such crafting positively draws on a shared predicament. Referring to their plight in conversation with fellow villagers, landless villagers mobilize common pasts and outlooks. This sense of being in it together, with the sole difference being whether the crisis has already arrived, underpins tolerance with respect to what formally are illegal settlement patterns.

The polity also appears in formal politics. Consider Partha Neogi: a short-tempered man, in his forties when we first met, he expresses his worries in a voice hoarse from smoking and talking loudly. He has seen many floods, and has had to shift his house several times. He has grown all too familiar with *bhangon*. Our conversations are interjected by rants against politicians and their carelessness. All are letting his village down, he continues to stress; the future is bleak. But his exclamations take on an eerie tone as he is a politician himself. Making a living for many years in a shack by the road, he has devoted much of his energy to working for improved embankment maintenance in a bid to safeguard his village from submergence.

Together with others, he has tried to push for consensus, and for what he called a "master plan." I will return to this idea, cohering around the lavish application of concrete, and its implications in chapter 6, when I discuss the politics of embankments.

His position as elected *panchayat* (village council) member seemed promising for a while. In this function, he had made formal requests to the state authorities, had leveled pressure in the corridors of power, and, repeatedly, spoken to external visitors, from engineers and media people to social scientists. None of this bore the fruits he sought. Botkhali continues to erode, its disappearance imminent. This is not the place to discuss tactical efficacy, the workings of the state, or abandonment. Instead, I rather want to suggest that in attempting to put pressure by mobilizing numbers and shared experiences and trajectories, he contributes to fostering a polity in time. Put differently, his working and tweaking of formal politics helps to bring this polity, and a collective experience, into being.

As I write this, Partha Neogi still sticks to Botkhali, waiting it out. But his hut has shifted once again, now to a more respectable place off the street. This scenario highlights another set of fuzzy edges of this polity. Much like the morphing shapes of the shore, or the outer line of homes and people immediately in the grip of the estuarine waters, the polity transforms continuously. Relentlessly, new people enter its fold, while others opt out, putting distance between themselves and eroding shores ripe with displacements. On one level, this resonates with disasters' descent into the ordinary, as people continue to cope (V. Das 2006). On another level, however, it signals the ongoing nature of the distributed disaster and, more importantly, its denial of a somewhat stable polity beyond the family and kin. All the while, as I have noted, in objective terms, large numbers of victims emerge in time. Their articulation as such is impeded both by the distributed nature of coastal erosion and by the fuzzy modes of finding what would locally count as more respectable ways of living, i.e., residing in a house on secure land.

A few miles away—in Gangasagar Colony, where Nur has now settled, sleeping badly during nightly spring tides—a much more robust, spatially fixed polity emerged. People living here look back on pasts marked by displacement at the hand of the voracious river. In these oldest parts of the colony, people share this trajectory, and they conveniently refer to once cohabited lands, to places and mutual acquaintances and kin. Yet, this is no unified or leveled polity. It is doubtful, of course, whether such a thing ever exists in the wake of disasters. Given varied vulnerabilities and the fraught character of rehabilitation in much of the world, rebuilding measures tend to be uneven, frequently building on pre-existing inequalities or cementing new ones. The "brotherhood in pain" (Oliver-Smith 1999a) dissolves into power-laden inequalities, at times even culminating in ghettoization (Simpson 2014). This is not only a matter of access and power, it also flows from the experimental character of implementing or devising schemes, of making do, in humanitarian contexts. Local administrations translate schemes into reality in particular ways, belying notions of uniform, impersonal statecraft (A. Gupta 2012; Spencer 2007; N. Mathur 2015). As a result, reconstruction, rehabilitation, and relocation on Sagar's morphing coasts are marked by an unevenness, which I explore in detail in chapter 6. The texture of coastal erosion, its distributed nature, enhances these dynamics.

This is relevant for studying rehabilitation. However, disasters become what they are only with time, involving what is often called "the wake" into thinking the event or process. This is also relevant for tracing coastal erosions and theorizing distributed disasters. Being distributed in time and space, state and developmental procedures addressing these processes vary considerably—not only in the way in which different local administrations handle challenges, but also in how these very local administrations address challenges across time. The polity on and off the embankment Nur now calls home is characterized by such variations. Off the embankment, people reside benefiting from full settlement packages: land, titles, and frequently also state-sponsored concrete houses. On top of the embankment, Nur and his neighbors continue to live without land and titles, in the proximity of murky waters.

With the contours of the polity here rather set, and the polity itself in the meantime rather stable, deep rifts prevail. Given shifts in governance priorities, and, arguably, the ends to available lands, living and coping with *bhangon* take different forms. In going through uneven resettlement policies, residents of Colony Para arguably experience different disasters. In other words, the distributed nature of coastal erosion is complicated by the shifts in governance across time and space.

Over the years of our friendship, Nur became more and more accepted as part of the polity. Tending to the dike during spring tides might have helped, as did referring to a shared history and a common predicament with his neighbors when engaging with bureaucrats. His inclusion materialized in the form of receiving funds for a concrete house under the state poverty alleviation scheme Indira Awas Yojana (IAY). The paperwork it involved gave him a sanctioned presence and de facto settlement from the bank, the second-best option below proper ownership titles to the land he claims.

Navigating Distributed Disasters

In this chapter, I have followed Nur and a few of his fellow islanders, exploring narrations, unpacking stories, and witnessing freshly displaced people in order to understand what is locally considered the most pressing issue. On these lands regularly ravaged by storms and surges, normalized landscape transformations appear as the real peril. Devastating lives over the course of years or decades and disrupting social-ecological systems, *bhangon* appears to islanders as *the* disaster.

Islanders' past and present experiences reveal that coastal erosion is a moving process that coheres around interruptions and intensifications, always combated by tense routines. Coastal erosion is a disaster spread out across time and space, distributed within the lives of affected people along seasonal rhythms. The notion of distributed disasters is attuned to this complex texture, while staying clear of deeming coastal erosion to be simply unbound or general. As this chapter has unveiled, thinking erosion as a distributed disaster embraces various degrees and timings of going through catastrophic devastations. Thinking it as a process sustained by quasi-events and interspersed with repeat inflections or culminations into very eventful aggravations has the advantage of theorizing the temporal, spatial, and social spread of dislocations well beyond everyday crises or chronicity. Meanwhile, it also makes room for noticing moments of heightened despair in such chronic conditions. It goes a long way, I suggest, to account for how the slow violence of environmental degradation adds onto, and is being lived through, the structural violence of normalized neglect, poverty, and environmental degradation.

I have demonstrated that coastal erosion brings forth specific challenges. It is not so much that they befall society, seemingly out of nowhere, out of sync, or as a surprise. *Bhangon* is well-known, and clearly anticipated. Nor does it hit a given society at once, spelling out large-scale devastations, fatalities, or massive economic losses. Quasi-events give way to small-scale destructions of plots so small, homes so poor, and villages obviously so irrelevant to schemes of regional politics that they appear inconsequential by most standards. Yet, homes are unmade, and lives marred by this process—often, as I have shown, several times all over again. Coastal erosion thus appears as a rolling disaster, encroaching on homesteads and families, shifting the zones of devastation as it moves along. At present, inland, always inland.

I have also shown coastal erosion to straddle individual hardship and collective predicament. As people are caught up by the sea one after the other across spread-out coastlines; as they are displaced by waters frequently five to seven times; and as erosion is seen at work in cyclical times of heightened danger, specific collective experiences and a shared perception of threat emerge. Repetition, seasonality, and the retrospect knowledge of having gone through it all gives rise, I have suggested, to collectivities and polities. Finally, the widely told story of the tree and the children gives meaning and a framework to think and narrate *bhangon* by rendering it spectacular. The individual tragedy then comes to be a moment to bolster a sense of misfortune that is otherwise difficult to pronounce.

Tracing how people navigate distributed disasters, and what kind of stories they tell about them, is instructive far beyond Bengal's delta front, and beyond coastal lifeworlds on our injured and overheating planet (Eriksen 2016). Distributed disasters oscillate between individual experience and collective predicament. They alternate, harming quality of life, with periods of relative quiet swatted away by devastation and turmoil. This aids our general comprehension of life amid planetary injury. What I have uncovered here is not simply ruination or a process of being "left behind by their own countries" (Latour 2018, 6). I have called attention to displacement, that is, an undoing of place and the unraveling of lives that is as predictable as it is unstoppable, running at its own pace and rhythms. Similar applies, I would propose, in a range of deteriorating landscapes. The process of places losing their hospitability before they fade or disappear affects today large populations living through morphing landscapes along the edges of deserts, erratic rivers, and retreating glaciers, or by the side of industrial compounds.

In being disastrous and disruptive, such drawn-out yet spiking predicaments require care and assistance beyond what customary disaster programs have to offer. In fact, this chapter has argued that such processes effectively remain below the radar of humanitarian interventions also seeking to frame disasters. My account of coastal erosion as distributed disaster, therefore, is not meant to be an academic exercise refining the conceptual apparatus of disaster studies. It can also help reorient disaster governance and humanitarian assistance on these shores and beyond.

Chapter 4

Seeking Shelter in the Mouth of the River

"Where to go?" The question surfaces in almost all of the drawn-out conversations and encounters that shape this book. Sunil Maiti would put it to me as he combed through what was left of his home, stranded on the wrong side of the embankment and now within reach of every tide. "Where to go?" Binodini Das would ask me, squatting in front of her hut when it still sat on the outer embankment of Ghoramara. The hut is now long gone, the shore having moved on since, and I have lost trace of Binodini Das. The question echoes in my mind as I wonder where she went and how. But it is one not only people in immediate distress reflect upon. It pops up in media accounts of displacement in the Bengal delta (see, e.g., Saha 2009). And it comes up among people out of coastal erosion's immediate reach. At one point or another, all my interlocutors in Colony Para would pose the question. Considered to be the lucky ones by so many others struggling in vain to secure resettlement, they would still voice it. "Where to go?" It accounts for tough choices—past, present, or future.

The short phrase does many things at once. It is a rhetorical question, hinting that there is essentially nowhere to go. It is a sigh, the voicing of distress. It is a framework to assess, and live with, exhaustion. In relating to an elsewhere, or its absence, the question, paradoxically, renders the here and now inescapable. In this view, it is a tool of endurance.

Unsettling the Refugee Imaginary

In this chapter, I take this question as a starting point and explore its contexts and ramifications. Along with an account of the suffering the question entails, I ask for the specific forms and modes of endurance it enables. This compels me to consider the notion that there is nowhere else to go—that is, to explore the past and geographical imagination as a horizon of contemporary anticipations. Once, the coastal seams were seen to be ultimate frontiers and remote spaces of opportunity allowing people to exit more crowded parts of southern Bengal. These opportunities have been used, space is shrinking, and land is, in itself, coming to an end as the economic propensity of agriculture declines, and therefore parts of the land's potentialities fade (Li 2014). Now they appear as endpoints of journeys, as dead ends, and even as entrapments. Exploring this double meaning of land—as an exit and a dead end—allows me to trace how futures are being wrought as people bet on the means at hand on this shrinking and besieged island.

Thinking with the phrase "Where to go?" also fosters critical engagement with the literature on environmental displacement or, more broadly, on the nexus of environment and migration. The scholarship heavily emphasizes long-term forms of mobilities, arguably echoing societal anxieties of refugees swamping affluent cities or continents (Hartmann 2010). Recent writing on the human condition highlights the pervasiveness of mobility (Urry 2007). Research on the nexus of environmental change and (forced) migration tells us that being pushed around by shifting weather patterns or major or minor upheavals of the earth is more of a norm, rather than an exception, in the *longue durée* of human existence (McLeman and Gemenne 2018).

If mobility is pervasive and migration increasingly global, most displaced persons still remain in their home countries (IDMC 2021). Many actually are unable to make a move at all. Only recently, scholars have begun addressing immobility amid displacement and turmoil. The language of "trapped populations" being "too poor to finance migration" is instructive (Black and Collyer 2014, 290; cf. L. M. Hunter 2005; Lübken 2012), and resonates with the question that guides me here. Inspired by such approaches, recent ethnographical work turns to ways in which people situate themselves within, and not in spite of, mobility. Making homes and cultivating life, then, emerges as a practice that accommodates both anchoring and mobility (Ahmed, Fortier, and Sheller 2003; Thiranagama 2007). Places appear rather like "crossings" within trajectories of living (Tweed 2006; Ingold 2000) than containers within which life occurs (Appadurai 1988; Malkki 1995).

In the Sundarbans, "Where to go?" seems to stretch beyond these explanations. I suggest it refers to ways of imagining an elsewhere that are informed by historical experience yet infused by apprehension of the future. Both the troublesome pasts and the narrow futures indicated by the question demand more critical work, more than merely tracing the certainly important dimension of "root causes" of disasters or displacement eagerly attended to by students of disasters (Wisner et al. 2004). It also requires tracing how the question shapes visions of surviving and living well amid adversities.

From its inception, forced migration studies placed heavy emphasis on the disruption of everyday life, and on attempts to mend or recreate it. In research spanning decades, Elisabeth Colson (1971, 2003) demonstrated how people in today's Zambia recreated everyday life after being relocated to make room for the hydropower dam at Kariba. In India, Amita Baviskar (2004) showed how Adivasi first fought displacement for the sake of hydropower and subsequently navigated resettlement. Both accounts cohere around disruption, or its looming threat, evoking forms of life firmly inscribed into place that require recalibration and re-anchoring after the event. To be sure, neither account denies mobility as a more general framework of existence, be it in the form of deep histories, of labor migration, or of activist travels. Through the lens of the dam, however, we get to know people whose roots were partly cut and who now struggle to lay new ones.

In this reading, I take the question "Where to go?" as an invitation to consider situated accounts of past and future horizons, and place them alongside geomorphological or political economy accounts. The aim of this chapter is to situate environmental displacement within problematic pasts so as to get a better grip on everyday strategies of going through it, as well as to problematize resettlement operations. I do so by engaging with what islanders call "life histories" (*jiboner itihas*). This chapter takes issue with an assumption that is at once implicit in and fundamental to much of contemporary debate on forced migration. Much of it coheres around the notion that prior to some external shock, be it a disaster or a conflict that made people move out, they were firmly anchored in place. I call this frame "the refugee imaginary," playing on its connotation of the passive refugee and a singular experience that marks a supposed stark contrast to both life prior and life after (if managed well).

If going through coastal erosion is, as I showed in chapter 3, a distributed experience that comes in rhythms, its prehistories have not been serene either. Long before the estuary encroached on them, islanders' lives were shaped by precarity, ousting, and flight. Extending my analysis of environmental displacement as a recurring phenomenon, I show how it maps on deeper layers of enforced mobilities. I capture these resonances, and the ensuing sense of repetition, through the phrase "enigma of departure." Here I take inspiration from novelist V. S. Naipaul, who reflected on migrants who had settled in the UK after initial sojourns across the

Commonwealth. Naipaul (1987) captures this feeling of uprootedness by referring to the "enigma of arrival." Turning the phrase on its head and speaking of an enigma of departure moves beyond reading migration as a singular disruption. Simultaneously, it underlines how departing comes to be an asset as long as there is a place worth departing for.

Islanders imagine their lives as marked by recurring times of claustrophobia and distress culminating in involuntary forms of mobility. Displacement by salty waters and submerging lands ceases to be an entirely novel phenomenon. Going through it seems to partly trigger rather well-known experiences. Troubles emerge as manifold, and displacement takes several forms, straining life in most places, yet also providing space for entrepreneurial successes and feats of survival.

This chapter situates the more recent and ongoing experiences of environmental displacement within broader past trajectories. I uncover what I call "circuits of displacement and emplacement" that guide the way islanders along the seams of the land situate themselves. Yet, this is not an exercise in peeling off layers one by one, diving deeper into layers of the past as if they were neatly mapped onto one another. Instead, I trace the way in which pasts are molded and remolded in the practice of remembering. This not only means that the past is accessed, read, and performed from a specific vantage point. In a dialectic twist, it also works the other way around, as specific ways of imagining the past inform future outlooks.

I begin with a story explaining Ghoramara's name.

What's in a Name

Anil Patro, bent by age, greets us. There is little to do in the time between lunch and his evening duties in the village shrine close by. He has a mat rolled out on the cool ground by the side of the pond, where we sit and talk. The mud house hardly resembles the ephemeral type of housing associated with the refugee experience or the standardized buildings making up resettlement colonies across the world. There is a hint of modular living in his alley with standard plot sizes and parallel lines. Nevertheless, a typical village feel is present: thatched mud houses hidden under fruit trees and overlooking ponds teeming with fish, tiny vegetable gardens, and paddy fields.

Soon the conversation turns to the past, how his family made it onto Ghoramara, the tiny island he calls home, how they lived before *bhangon* knocked at their door. These are pasts marked by the mobility of both people and lands, and, most importantly, by cascading troubles, by tightening prospects and enlarging riches both waxing and waning across time. At one point, he asks whether I know how the island of Ghoramara got its name. In Bengali, Ghoramara means "the dead horse." I have heard several iterations of the story but remain curious and encourage him to share his account. Listening to his version, I admire not only this gifted storyteller's eye for detail and cadence but also how this storyline captures how early settler life, with its threats and openings, is remembered.

Anil's story went like this:

How Ghoramara got its name? That is a long story. When Ghoramara got its name, Sagar Island did not yet exist. Sagar Island was water, only water. Ghoramara emerged before [*upardhan hayechilo*]—therefore the first and oldest area [*acal*] in Sagar is Ghoramara. Only later did Sagar arise [*utphanna hayechilo*]. In those days, Ghoramara was for nobody, nobody lived there. Only jungle. Into this jungle, one British man came to take a walk.

Looking here and there, he arrived with a boat and, roaming on the island, he saw the jungle. He lived in a boat on the shore and had seen the entire jungle but he had not settled down. He thought, when the jungle would be cut and some people would be brought and settled here [*kicchu praja-basati karle*], then people could live here—such a wealthy sahib he was. Cutting this jungle, then, he let some subjects live here as an experiment. He let Adivasi cut the jungle and told them: this is yours forever; as a reward for your struggle against the dangerous animals, this is yours. In this way, the cutting of the whole jungle was going on: he gave money, but not as wages according to the work done, but gave it in this way [in the form of land].

From there on, settlers came to the land, and embankments and all that were built. But since it is an area ruled by the tides [*joyar thana*] and floods in the interior parts too, embankments had to be built everywhere. In this way, the people undertook his work. He also had an agent and an office—in this office, his agent did all that was necessary to rule. His sahib came monthly, every fifteen days or fortnightly with money, distributed some, and sailed away again. How much money was spent within the time [of his absence], to whom it belongs, to whom it was to be given and how much—this he calculated in his absence. After fifteen days, he came again to give money to those who made it, to learn from papers whom he was to give money to and to give it to them. In this way, the work began and went on.

In the meantime, the sahib and his younger brother came together on horses. That is, they took the boat, reached the shore with the boat, entered [the island] on foot, and then rode their horses along the entire embankments. On the top of the embankments, they were riding their horses and watched from up there if the people working in the interior did their work properly, if they cut everything, and to speak with the people.

In these times, in the jungle there was a tiger. The tiger approached the younger brother and took him with him—the younger brother. As the younger brother had vanished into the jungle and suffered, his dead horse was still there. He [the older brother] had gone down and watched out, and only the horse was there. Where did the brother go? Inspecting the area, he only found the dead horse, but not the brother. He realized that the younger brother had been taken by the tiger, that the tiger had picked him up with a loud roar and had vanished; his turban and all that lay spread out, and blood, a lot of blood from torn wounds had fallen onto the ground. On seeing the blood, the sahib thought to himself, *I will not leave from here together with him* [his younger brother]. *No. And this estate I will not keep*, he thought. In this state of despair, he returned to his house. Back there, he said to himself, after having thought about it long, on this night's morning, *I will give the land free for ninety-nine years to the first person whose face I shall see. That is, I will give it to him with papers.* When the papers were written, he looked in small shops for boatmen who would transport his freight. In this way, after he made some arrangements, he returned to his house. He did not enter the jungle again, since the brother had disappeared because of the horse. As he returned to his house, the neighboring house was the house of Perimon Mukherjee of Bardhaman, who also lived in Bardhaman [a district north of Calcutta].

At that time, in early morning, he met him carrying a small jug with a narrow neck. As the sahib got to see him, he gave the papers and the estate to Perimon for free. That king [*raja*] let the whole forest in this estate be cut, the whole population and all that [*praja-traja*] settle down and all that.

From the horse that was killed and eaten by the tiger, from this the name Ghoramara [*lit.* dead horse] originates.

Anil's rendition sheds light on how islanders and visitors continue to conceive this part of the coastal archipelago. Alongside personal tragedy, we are introduced to putrid jungles at the end of the world, and windows of

opportunity among rural paupers willing settlements into life by enduring the vicissitudes of the frontier. I will unfold these frames in order to
flesh out the horizons of immobility and the powers of the question that
instigated this chapter—where to go?

But before I do so, I need to unpack a further frame of situating the
island.

A Horse Sacrifice Gone Wrong

On another occasion, I joined Anil as he held a public screening of
one particular episode of the hugely popular TV show *Ramayana*. It was
the night of Shivaratri, a night to celebrate Shiva. Anil thought it fit to
do so by inviting family and friends to watch the show on a TV set carried to one of the two village temples for the occasion. Watching this
show was not merely entertainment but a form of service, as has been
noted in other South Asian contexts (Lutgendorf 2006). Here I am bypassing a discussion of media as ritual, instead turning to the show for the
way it positions Sagar on the pane of mainstream Hindu cosmology.

The show provides a close, if reductive, reading of those sections of
standard Ramayanas detailing the descent of Ganga from heaven. I para-
phrase the story (for extended versions, see for instance, Valmiki 2005).
It goes like this:

> King Sagar arranges for the sacrificing of a horse. The sacrifice involves
> allowing a horse to roam freely. Whoever stops the horse challenges
> the authority of the sacrificing king. His army—who are following the
> horse—will open fire, thus testing and extending the outer limits of
> the king's dominion. At one point, the god Indra steals the sacrificial
> horse unnoticed, disrupting the sacrifice and humiliating the king. King
> Sagar's sixty thousand sons then agree to search for it. After searching
> the whole world in vain, they eventually turn to the underworld. Tear-
> ing open the ground, they encounter monsters, bringing chaos to the
> dominions of snakes and other underworldly entities. They finally find
> the horse in the deepest corner of the Eastern netherworld as it grazes
> calmly in front of a hermitage to which it was tied. Infuriated, the sixty
> thousand sons confront the yogi residing in the cabin, Kapil Muni,
> about why he stole and hid the horse. Kapil Muni is offended by the
> lack of respect demonstrated by the princes. Insult turns to rage, and—
> swelling up the enormous power accrued by his penances in the form of
> inner fire—he burns the princes. All that remains is a pile of ash.

Soon, King Sagar begins to worry. No news has come in, and the sacrifice is still incomplete. Anshuman, his grandson, volunteers to look after the sixty thousand. Following the trail of destruction, he finally makes it to Kapil Muni's cave. Approaching him respectfully, the yogi tells him about the fate of the princes, much to Anshuman's horror. His relatives are dead and have died without final ablutions, which bars them from entry to heaven. After Anshuman pleads with Kapil Muni, the yogi tells him that only the purifying powers of Ganga could wash off the sins of the princes. Anshuman decides that, after returning the horse to his grandfather, he should seek to convince the gods to send Ganga all the way to the pile of ashes still sitting in front of Kapil's hermitage and have them released. Gods of the Hindu pantheon tend to bestow wishes only to those who demonstrate their devotion through penance and asceticism. So, Anshuman sets out on a life of penance, followed first by his son and, after him, his grandson. After three generations, the gods finally give in. Brahma grants the wish to Anshuman's grandson, Bhagirath, to lead Ganga to the ashes of his ancestors.

Brahma convinces Ganga to leave her heavenly abode. Shiva agrees to cushion her downfall in his matted hair, so Ganga descends through the Himalayas. Bhagirath welcomes her, showing the way across the plains and down to the cave. Wherever she arrives, her arrival brings joy as she provides purification and bestows fertility and abundance to all along the way. Bhagirath finally takes her to the cave in the netherworld, where she washes the ashes clean, setting the princes free, and empties into the sea.

Situating Sagar in the worlds of scripture and folklore, the narrative also provides for theological and cosmological clues that prove significant for gauging how islanders engage with and relate to their home island.

Hindus tend to not only know the narrative by heart, but also to read it literally. To them, the river is a watery body of divinity, emerging from Shiva's locks and providing nourishment and purification across the North Indian plains before eventually bringing its journey to a conclusion in what used to be Kapil Muni's cave deep in the netherworld. I will return to this trope in my discussion of narratives of blame and protection (chapter 7). Suffice for now that the literal reading of the narrative suggests that Sagar marks the deepest place on earth, a former cave where the powerful yogi Kapil Muni resides.

Such visions animate pilgrimage undertaken by hundreds of thousands of Hindus every year, worshiping river and yogi on Sagar's southern sea-facing shore. They also underpin pilgrimage infrastructures and traffic between island and mainland, which, in turn, shape the way the island is related to by all people involved, from islanders to those just passing by. The figure of fluid divinity continues to be woven into the palimpsest the island is. I will return to these dimensions in chapter 5, where I detail how Hindu-inflected visions of development of the pilgrimage center inform resettlement experiences. For now, I merely want to emphasize that these two stories introduce the islands as remote spaces of violence. That is, they render the islands endpoints of journeys, cul-de-sacs that protagonists leave only harmed and by way of backing out.

The trope of violence and death continues to inform pilgrimage visions. Pilgrimage almanacs speak of the risks of traveling to southern shores (Bhaṭṭācārya 1976). Until quite recently, the dangers seemed manifold. Man-eating tigers lurked in dense forests that pilgrims had to cross on foot. Powerful currents, thunderstorms, and cyclones hurt or killed those betting on seafaring. Pirates threatened to sink boats and drown men. In the recent past, boat accidents have caused injury. However viewed, pilgrimage narratives introduce the islands as remote and risky ends of the world.

Both accounts—Ramayana's positioning of Sagar on the map of sacred India and Anil's version of how Ghoramara got its name—are also connected through the image of the horse. Ghoramara literally translates into "dead horse," tying the island to the bloody end of that white man's joyride. Meanwhile, according to the Ramayana, it was the mischievous doings involving a horse, and their deadly consequences, that invited the river Ganga down to earth before setting afoot a continuous stream of pilgrims to this outpost. The connection is admittedly loose, but tracing it allows me to achieve several things at once. Considering the horse enables me to tease out how Sagar and its environs were thought of as spaces beyond the control of political dominions, and, by extension, unmarked by violence rooted in political and economic strife. It allows me to situate Sagar on Hindu sacral geography, to pinpoint how the island straddles divergent cosmologies, or better: how people navigate Sagar straddling divergent cosmologies.

Recall how Anil began his account of Ghoramara's name by stating that the incidents reach back to a time when Sagar had not even risen from the waters. This framing indicates not only a vision of fluctuating, young, and volatile lands, it also enters a tension with mainstream Hindu accounts. It implicitly challenges the antiquity of Sagar, and positions Sagar as an island arising out of the floods, not a cave rinsed by celestial waters. Since Anil was also arranging for a public screening of the *Rama-yana,* and firmly maintaining at other times that Sagar marks the deep-est place on earth, his take was an ambiguous one, mobilizing separate cosmologies all at once. Anil, of course, is not alone in accommodating such ambiguities.

Take Shankar, for example. On a different day, I find him stretched out smoking on the planks that serve as furniture in his simple roadside restaurant. He has that characteristic grin plastered across his face. As he is a devout man who makes a living from serving rice and fish to pil-grims passing by on their way from the main road to the main temple, his amusement was unsettling. Pilgrims often believe, he says, that the island is under water most of the year, sunken below the surface of the sea only to rise up every year for the festival—the Gangasagar Mela, which attracts masses from across the world. As we speak, the heat of summer pours through the windows, and the last mela, falling into the damp of mid-January, is long over. His grin seems to convey that if the pilgrims were right, we would be underwater by now.

It is tempting to read a sinking and rising island in terms of theo-logical metaphors, stating that they articulate the idea that pilgrimage spots are crossings, *tirtha,* that allow for access to the worlds of divinity within spatial and temporal frames. It also is tempting to read this idea as a marketing trick or an act of romanticizing that works by emphasiz-ing the exclusive qualities of this place to devout Hindus. Likewise, it can be read as yet another variation on the theme of sunken cities and drowned worlds that haunt the imagination (Ramaswamy 2005). While all these readings might be useful to account for the tale of the sunken island, I am struck most by Shankar's grin and the delight he conveyed as he told the story of his fellow believers' rather odd belief of what the island was and where it lay. Perhaps he took delight from knowing better and from pin-pointing a ludicrous belief, but the smile also hinted at the rather bizarre fate of positioning this island and its encompassing waterscape on two very different planes of existence and of holding fast to both. For he too understood the island to be simultaneously a rather ordinary island *and* Kapil's cave. Disagreeing with the pilgrims' belief of a submerging and uprising island, forever in suspense, he shared with them, and with most islanders, the idea that they were dwelling in the deepest place on earth.

As to the tales of horses and violence, neither has an anarchist plot. In both cases, authority is upheld: in the form of a man of upper-caste origin in one, as his full name implies, and in the form of a short-tempered, invincible yogi in the other. Still, ownership could slip with the snap of a finger, and state authorities could be overwhelmed by some raw powers, with lesser stately people taking over or becoming someone themselves.

Both stories also draw upon the opening up of spaces as a means of bringing about possibilities. The story of Ganga's descent speaks of streams that fertilized the plains and made the delta out of what was a cave hidden deep in the ground. In other words, both the thick greenery and the solace of the cave appear at once as spaces of death and of future prosperity, lying dormant and waiting to be realized. This version of how Ghoramara got its name invokes clearings—spaces where jungles are cut, opportunities seized, and fertile futures forged. This resonates with the ways islanders narrate their own or their recent ancestors' pasts. As I will show, the outer contours of life histories were marked by visions of a swamp that altered lives—in order to survive and achieve modest standards of living, to move out of prospects that are ever tightening through the twin processes of impoverishment and suffocating "tradition" that continue to define life in the rural plains.

In the story of the dead horse, we hear little of poor Bengalis colonizing the swamp—though I will add detail to this in the sections that follow—but enough to lay out the horizon of settlement. It was fueled by humans desperate enough to flee their former homes and to toil in a perilous jungle, and pioneering enough to do so with an eye on future bounties and the possibility of improving life quality through the amount of available space. This constellation of layered suffering forms, as I will argue, the background and recurring theme of the phrase "Where to go?" I identify three tropes of such suffering and how it drives people to up sticks. First and most prevalently, there is the trope of being ousted by land scarcity and poverty. The second refers to escaping disastrous events, while the third speaks of illegitimate love relations. As I will show in the next section, these three tropes overlap tightly in emphasizing distress and loss, where moving into risky terrain haunted by tigers and malaria appears as perhaps the only strategy to deal with these circumstances.

Putrid Jungles in the Mouth of the River

It looks as if this bit of the world had been left unfinished when land and sea were originally parted.

—Emily Eden, *Up the Country*

"The Sundarbans are the frontier where commerce and the wilderness look each other directly in the eye; that's exactly where the war between profit and Nature is fought," notes novelist Amitav Ghosh (2019, 8–9). Anil's iteration of the story of the dead horse focuses on that fight. What he fails to mention, and what forms an important backdrop, an integral piece of the puzzle, is that for centuries the entire marine traffic between Calcutta, the glistening capital of the British Empire's crown jewel, and the wider world navigated along the western edges of the Sundarbans. Sagar Island marked the eastern flank of the port channel, its lighthouse a sign of the city nearby. Yet the islands of Ghoramara and Lohachara were actually never named during this period, and it remains a little ambiguous as to why. Geographers date the former's birth between 1903 and 1904 (Nandy and Bandyopadhyay 2011, 810). From the deck of a ship, the contours of individual islands might have blurred into one another; or the sheer scale and terror wrapped up in perceptions of the neighboring jungle paled the desire to track individual islands.

When passing those lands onboard passenger ships, writers revel in getting close to land after long days of sailing on the ocean board. But what land it was! Visitors, such as Maria Graham, whose account of her years in India was a huge commercial success, notes (1812, 133):

> The water looked like thick mud, fitter to walk upon than to sail through. . . . Nothing can be more desolate than the entrance to the Hooghly. To the west frightful breakers extend as far as the eye can reach, you are surrounded by sharks and crocodiles; but on the east is a more horrible object, the black low island of Saugor. The very appearance of the dark jungle that covers it is terrific. You see that it must be a nest of serpents, and a den of tigers; but it is worse, it is the yearly scene of human sacrifice, which not all the vigilance of the British government can prevent. The temple is ruined, but the infatuated votaries of Kali plunge into the waves that separate the island from the continent, in the spot where the blood-stained fane once stood, and crowned with flowers and robed in scarlet, singing hymns to the goddess, they devote themselves to destruction; and he who reaches the opposite shore without being devoured by the sacred sharks, becomes a pariah, and regards himself as being detested by the gods. Possessed by this frenzy of superstition, mothers have thrown their infants into the jaws of the sea monsters, and furnished scenes too horrible for description; but the yearly assembly at Saugor is now attended by troops, in order to prevent these horrid practices, so that I believe there are now but few involuntary victims. As we advanced up the river, the breakers disappeared, the jungle grew higher and lighter, and we saw sometimes a pagoda or village between the trees.

In hyperbole characteristic of travel writing on Asia at the time, Graham highlights, in a condensed form, the themes through which Sagar was then known. In her account, both the thickening waters and the forested land cannot be trusted and swarm with deadly creatures. But more treacherous than the evils of nature in the form of alligators, tigers, or snakes seem to be the people sacrificing their daughters.[1] The "black low island of Saugor" appears as a darkness that one had to go past in order to reach the civilized premises of the capital (also see Arnold 2006, 71–73). Considering the jungle island's darkness and gloom, even the notorious "Black Hole of Calcutta" (Hutnyk 1996; Chatterjee 2012) appears as a relief and, potentially, a scene of enlightenment and refinement.

A few years later, Walter Hamilton (1815, 711–712) concluded his brief description of Sagar Island in his *East India Gazetteer* after detailing once again the shuddering practice of infanticide by noting that "on shore the jungles swarm with tigers of the largest and most ferocious sort, so that both elements are equally dangerous." Both descriptions serve as good examples of Victorian accounts of India's jungles, accounts that are, as Arnold (2006, 81) argues, replete with "multiple associations between physical harm and moral evil [in a landscape] that was perceived to be both heathen and deadly, as entangled with rank, miasmatic, over-fecund plant life as Hinduism appeared to teem with primitive beliefs and convoluted superstitions."

The taut association of jungles with tigers continues to be emblematic for the way the Sundarbans are envisaged in the region and across the globe. The danger of deadly reverence is specific to the Sagar. In some ways, it renders the island visible as a site of mass death and despair rooted, as it were, in a murderous sea.

---

1. Graham joined the ranks of visitors spreading the narrative of infanticide. By the time Graham passed the island, Sagar had risen to fame as a site of the ritual killing of girls. Some observers noted how Hindu men drowned girls in the waters in order to receive merit or to have them reborn in the more auspicious form of boys. Missionaries decried the wretchedness of the local faith supposedly motivating men in numbers to kill female family members. Following scandalizing reports, the practice was officially outlawed in 1802. The rumour of infanticide on Sagar's shores fed into the growing attention on the abuse of women in India. The ritual killing of girls was seen as another instance of the frenzy of an idolatry-ridden, despotic race, in need of pacification by patronizing colonizers. The vision of drowned girls signals how the "White Man's Burden" frequently involved what Gayatri Spivak (1988, 296) evocatively calls "white men saving brown women from brown men."

With the exception of juggernaut pilgrims, the jungles were depicted as empty. This is much in line with colonial and imperialist tendencies to willfully overlook, neglect, and stamp out human relations in landscapes targeted for development, thereby proclaiming *terra nullius* (Kapila 2022). Long before the advent of the English, however, settlers, traders, and petty chiefs traversed the maze of swamps and shifting islets to capitalize on the bounties of the Bengal delta and its marine waterways. Early settlements are subject of scholarly debate, such as whether Sagar at one point served as a pirate stronghold, close to maritime networks yet remote enough to disappear in creeks and jungles (Sarkar 2010), or whether it merely figured as a point of navigation en route to settlements upstream (Chattopadhyay 2020). Stories linger of pirates who raided the delta and used the intractable thickets of mangrove swamps to hide, with some even settling there. These violent, intransient pasts continue to be evoked by today's islanders when they relate to their remote islands hidden within a maze of waterways as a "territory of pirates," *magher moluk*. If slave raids may be a thing of the past, the capture of humans and their belongings on the swampy frontier persists (Cons 2021; Gupta and Sharma 2009). More than that, relating to the islands as a *magher moluk* implies the absence of tight state control and a lawlessness emerging in its interstices. As in much of the Sundarbans, the label "pirate's den" invokes spaces of unruliness and endemic violence (Jalais 2010b, 4). In this reading, the rugged terrain meets the ingenuity of law-bending individuals who know how to exploit it. It also calls attention to an undercurrent of violence that many islanders see as constitutive of present-day Sagar. By this I mean not only the violence of predatory extraction, instantiated by mis-development, corruption, or the disdain for islanders' survival at the hand of actors or institutions practicing state "uncare" (A. Gupta 2012). I also refer to islanders' assumptions that they themselves tend toward violence. To some, this inclination is borne from ecological conditions, namely the high degree of salt in the water and the air. To others, it comes from the vicissitudes of navigating disasters and coastal erosion.

Exiting Strife for Aladdin's Cave

Anil's father settled on Ghoramara during the heyday of British-led settlement operations, which were concentrated on turning the obnoxious wilderness "into a seat of plenty" (Huggins 1824, 3). These efforts went through a series of ups and downs, and were, from the very beginning, controversial. Events like the gifting of a whole estate after a white settler's brother was killed by a tiger never made it into archives, and proof remains hazy, but a range of other troubles that affected settlement did. Urban investors received a blow when the government monopolized salt manufacturing, cutting the second pillar of a number of settlement operations (Pargiter 1934, 340). As always, storms swept the island. One, the 1833 cyclone, brought Saugor Island Society—an investment corporation located at Mudpoint (today's Ghoramara) and Dobhlat, in the neighboring Colony Para—to its knees. After the storm, the society gave up, sold its leases, and disintegrated (Pargiter 1934, 338; O'Malley 1914, 240).

With an eye on massive casualties, the colonial government burdened investors with disaster-protection measures, such as dykes or artificial platforms that provided refuge during surges. Even if implementation was diluted by capitalists balking at costly measures, punitive fees threatened their investments. As a precursor to contemporary affairs, government bodies had disaster management on their agenda, yet largely ignored the threat of coastal erosion or submergence. William Hunter, the prolific and hugely influential writer, circulated the idea that the Sundarbans were the remnants of a much larger terrain gradually submerging into the ocean (W. W. Hunter 1875; Greenough 1998). Similarly, geomorphologists warned of the fundamental unsuitability of the delta's seams for human habitation (Mukerjee 1938). The landmass being very young, diking and draining would cut it off from tidal incursions carrying sediments, thus disrupting the natural trajectory of growth and keeping them perpetually below high tide level. On their respective terms, both interpretations deemed the coast an ephemeral, amphibious terrain and emphasized the hazardous dimensions of such an in-betweenness.

Irrespective of these concerns and backlashes, settlement drives provided opportunities for rural paupers hailing from Bengal's plains. Here, jungles were open for clearance, and homesteads waiting to be wrought, or so it seemed, from tidal swamps. They signaled a way out of the density and scarcity texturing Bengal's agrarian crisis.

If the British sought to unlock yet another "Aladdin's cave [ready] to despoil it of its riches" (Lahiri 1936, 39), landless farmers rather sought means to secure futures in the swamp. This at least is what life histories suggest. From what I hear, Anil's father, for instance, was much like one of the many colonists wrapped up in his story. The outer contours of his life history were marked by grinding poverty. His family too hailed from Medinipur, the district on the mainland on the other bank of the Hugli

River. The land his father was to inherit was insufficient, his prospects bleak. The already meager plots were to be split between brothers, not enough to sustain a family, let alone work toward any kind of comfortable life. Want is mirrored in his version, as in those of so many others who experienced similar fates. Dense villages, tiny fields, and crumbling houses all signal rural crisis as embedded in their existence.

Conflicts erupting in cramped homes add further strain. Many speak of being outmaneuvered by inheritance laws—which demanded the distribution of plots among brothers, fragmenting assets into miniscule plots, or favored the firstborn son at the expense of all others. Tensions ensue, flaring up into bitter fights. Many others speak of being weighed down by the expectations of family members, driven by norms of setting up a home. Thus, in addition to having no prospects, sons felt ousted by increasingly hostile family members, pushed into finding income and futures elsewhere. Here difficulty (*samasya*) indexes both: poverty and family quarrels. But beyond rather sparse allusions to conflictive dynamics, most islanders remained silent and demonstrated a reluctance to speak on these issues.

Anil's friend Baishatta Jana, however, was fairly explicit about domestic disputes. He told me about his father's trials in saving money to buy a small plot and build a home, how they fled this place after being attacked by thieves, and how they returned to his family's residence destitute:

> We returned to our birthplace and my father told them that we need a place for shelter . . . My father said, "Give us a shelter here and then we see what lies in our fate." They gave us a space to stay that was not even enough to build a small hut . . . We began living there, and in due course my father fell sick. When he fell sick, we had a cow, which we sold off for his treatment. I had an uncle [father's younger brother], he was a sly uncle [*bodmais kaka*]. He always looked out for opportunities. He said, "Give me the money and I will do what needs to be done." He took the money and never gave it back. He also began to occupy our space [*jayga*]. After my father passed away, I had no one because my brother had died shortly after. And the place belonged to my cheating uncle. What should I do? Where should I go? There was a boatman living our area, he told me "Come along." I was a bit hesitant at the beginning as I had never been outside the house [away from the family]. But who would feed me?

He stayed with the boatman and eventually married a woman from one of the coastal islands before making a home with her.

At this juncture, scarcity, conflict, and violence lucidly intersect. On one level, there is violence unleashed by dacoits, undoing poor peoples' efforts to settle and claim a territory as their own. On another, more importantly, there is violence within households—as with the cheating uncle adding injury to the hostility offered by the family, in the case of young Baishatta, who remembers, in the wake of his father's death, having few options left other than to escape from dismal conditions with a stranger.

The historical accord shows that these are not exceptions. During the final decades of colonial rule, grating poverty engulfed southern Bengal and continued unabated after independence. Development stagnated as capital was diverted into the coffers of colonial lords by way of heavy taxation (Birla 2009; S. Bose 1993, 2007). Local industries were outmaneuvered for the benefit of British industry and trade enterprises penetrating the Indian market (Pandey 2006). As a consequence, rural Bengal continued to depend on agriculture, all while being embroiled in a global market economy on deficient terms. Mounting population density aggravated the situation. The majority of farmers in southern Bengal at the time were marginal landholders, with plots rarely sufficient to feed growing families (S. Bose 1993). Sharecroppers, on the other hand, often the only alternative for people disfavored by inheritance, were bound to deliver the majority of the harvests they produced to landowners, thus being kept poor and, frequently, knee-deep in debt. Among farmers, unrest emerged in colonial and postcolonial times, contributing critically to the rise of communist and Maoist movements that were to shape politics in West Bengal for decades to come (Dhanagare 1976; Mallick 1993; Basu and Majumder 2013). Below the level of agitations and party politics, however, people struggled on a daily basis, seeking ways to make ends meet under deteriorating circumstances.

Paralleling the better-documented flow of the poor who swelled the ranks of an urban working class (see Chakrabarty 2000), another stream fueled the colonization of tidal swamps at the delta's seams. As people sought ways out of caustic circumstances, and with lands settled thoroughly on the mainland, rumors about the possibility of acquiring land by settling jungles on one of the islands at the confluence of river and sea began to circulate. None of this was entirely new. Cutting and draining jungles to establish new settlements, and thus to move into the forest frontier, had been a widespread strategy for a long time (I. Iqbal 2010;

Blair 1990). It encapsulated visions of the good life, undergirded by kinship norms, where the settlement of new land and the extension of family networks seemed desirable. Driving settlements toward the fringes of the forest also reduced vulnerability to environmental hazards—spreading plots and homesteads across the terrain meant gaining access to a wide range of soil qualities and land formations, which in turn reduced the likelihood of all plots being hit by droughts or deluges.

Relying on hearsay or on existing networks of kin or friendship, and possibly also—although this never figured in conversations—on middlemen luring paupers into the jungle for the benefit of rich investors, people took to the coastal swamps in great numbers. These experiences serve as yet another indicator of the limits of distinguishing between voluntary and involuntary forms of mobility or migration. Does the escape from poverty, drudgery, and severely limited outlooks within claustrophobic conditions qualify as voluntary migration? But can we speak of involuntary forms of migration in situations where notions of the good life and kinship expectations involve the clearing of new ground and setting up homes from scratch? And how do both figure within historically deep strategies of reducing vulnerability to environmental risks by extending holdings widely across space?

Furthermore, these life histories substantiate that the experience of distress and displacement linked with *bhangon* was not singular. Islanders undoubtedly look back on troublesome pasts, evidenced by compromised living conditions and the disappearance of land or future prospects. This is not to suggest that environmental displacement is more of the same. In fact, this book argues that coastal erosion involves specific experiences and presents rather unique challenges. Yet, the experience of coastal erosion obviously maps onto deeper pasts of ousting and flight, or, in more chronic cases, of gradual impoverishment and despair.

Reading the coastal seams as Aladdin's cave, as settlement reports suggest, resonates with Hindu scriptures positioning Sagar as Kapil Muni's cave. While the imagery of Aladdin's cave plays on treasures and magic, more than on power and divine rage, both invoke a hidden gem, the endpoint of a search. These readings resonate with the way Anil and many others remember their parents'—or indeed their own—arrival at the swamp. Perhaps even more so than Aladdin's tale indicates at first sight, these were also blind alleys. Sagar's sixty thousand sons never made it out of the cave, much like the descendants of poor migrants now struggling to see a place to move out to. They are trapped, or so they envision themselves, by the fact that land has come to an end. As the lands have now been submerged by water, so have the chances of a dignified life in the lands they escaped to just a few generations ago.

It is no coincidence, I suggest, that as Baishatta recounted his life history to me, he also pondered, "Where should I go?" By posing this pervasive question, he does not put its urgency into perspective, I suggest, but rather hints at structural similarities across the different cases of involuntary mobility that are spurred into motion by *bhangon*. Such structural similarities also shaped future outlooks, as the phrase indicates. Baishatta and many others knew of the challenging conditions involved in making a living elsewhere beyond their fragile island homes. Men and women came back from extended periods on the trails of circular labor migrations, which take men to construction sites in Orissa or factories in Kerala and women more often to middle-class homes of Kolkata (Ray and Qayum 2009). Both ventures bring stories of conflict and overcrowding. Many encounter middlemen cheating them of their wages or superiors treating them badly in situations where they are unprotected by law or existing social networks. Others shared the plight of impoverished people living on the streets or in the cramped poor quarters of cities, without plots or neighbors to fall back upon. In such conversations, a sense of being trapped framed the present as well as the future, in ways that resonated with not-too-distant pasts. Imagining futures elsewhere beyond the delta and its hazards in many cases became constricted by the grinding poverty and marginalization they or their forefathers had escaped from not long ago.

Disruptions

We put Kasem Sikhari's life history together through a series of conversations that were animated by the interventions of two of his daughters-in-law eager to tell the story to the rolling tape recorder. It was thus a coproduction that signals the collective production and the elasticity of memories (Connerton 1989, 2008; Halbwachs 1992). Much of what was recollected by people partaking in the conversation took place on Sagar itself. Such an account may seem an unusual addition to those presented in this chapter. But, on closer inspection, Kasem's experience is a salient example of how events like floods and famines bring conditions to the boil, how the drama and visceral stupor inherent in disasters entangle themselves within life stories.

Kasem's paternal grandfather, he tells me, moved onto the island from neighboring Medinipur. He too escaped dwindling prospects as soon as he could. Offered land in exchange for his services as an armed guard to a wealthy landlord, he held land on both the mainland and the island. However, after a while, poverty hit hard again. His family's modest wealth crumbled with the coming and going of the flood—the Red Flood, that is: the surge accompanying the 1942 Medinipur cyclone. In his story, the force of the storm and the surge receded behind the troubling aftermath. Hunger prevailed, and property relations were reshuffled. At one point, his sons added to the story: "For five kilos of rice, ten kilos of rice or a mound of rice, he subsequently sold the whole of their twenty *bighas* of land in order to feed his family. People had nothing to eat. . . . In those days, human suffering was so intense that starving people collapsed on the street, dying. The government did not help."

Most narratives of the Red Flood indicate entanglements of social suffering and reshuffling of property relations, rendering it an instance of disaster capitalism (Sainath 1996). None I encountered, however, were so explicit on distress sales and involuntary forms of mobility. But I am more intrigued here by the last two sentences. The exact phrasing—"people collapsed on the street, dying"—resonates with accounts of the Bengal famine of 1943. Indeed, the figure of hollowed paupers roaming the streets of Calcutta or other southern Bengal cities in a desperate search for food is somewhat emblematic of life during this famine. Contemporary sources speak of living corpses dying on the street.

More remarkable, for my purposes here, is that both events appear collapsed into one in memory—as they did to many, as I have suggested above. This duo of events, taken together, merge drawn-out structural processes into event-like occurrences. This speaks volumes about the demands of remembering, of the relative accessibility of certain frames of remembering in place of others. But if disastrous events eclipse structural, slow processes of disenfranchisement and impoverishment, as the reading of Kasem Sikhari's life history along all the others I have presented here suggests, this may also illustrate another facet of what I have called success stories. That is, prioritizing disasters—be it the flood or the famine—makes room for rescuing one's life story from being a trajectory of failure or loss. Through the prism of the social disaster, it is not mismanagement of household resources, the pitfalls of family sizes, or the poverty that people failed to navigate, but a larger-than-life occurrence that engulfed society.

In due course, Anil Patro's and Kasem Sikhari's life histories became entangled with an emerging disaster governance. Sagar certainly became the focal point of experiments in governing a cyclone zone in terms of improvement and the naturalization of control (Kingsbury 2015; Bandyopadhyay 1997). These experiments cohered around embankments, whose maintenance had initially been put on the shoulders of investors before the state ultimately took over.[2] But further traces were also left. Close to Anil's former home, on Ghoramara, a platform from mud was raised. Some fifteen meters above the ground, it provided a safe haven for islanders expecting a surge. In an otherwise tremendously flat landscape, today the platform provides for a landmark. Villagers nowadays largely prefer to seek refuge in one of the islet's few concrete buildings, though some do head up this mound and weather the storms up here.

As a matter of fact, the platform was erected at a central location, much like the ordinances suggested, and is adjacent to several government institutions, such as the panchayat office, the school, and the market. This not only speaks of the trend of infrastructures to condense and overlay, but also signals the relevance of disaster mitigation for the governance of populations on terms of what might be called a disaster biopolitics (Marchezini 2015; S. Sharma 2001). Interventions such as these emblazon long-settled parts of the island with the imprints of an—at least officially—bygone empire.

This is not to say that all was struggle and pain. Listening carefully, and returning to critical events in extended conversations, also laid bare moments of joy and achievement. To them, and to their role within life histories of shifts, I will now turn.

---

2. Leaseholders—that is, investors who largely were absentee landlords—were burdened with mandatory clearly defined protection measures, such as constructing "a central place of refuge" in every estate "consisting of a tank surrounded by an embankment 16 feet high, that no habitation should ordinarily be built more than a mile from a place of refuge, and that embanked paths should be made connecting the places of refuge with the houses" (O'Malley 1914, 241). Unsurprisingly, such schemes hardly materialized to a full extent. Writing about Sagar, Ascoli (1921, 78) notes that "protective works were incomplete throughout; the water in the tanks was undrinkable; and no embanked paths had been constructed." Ascoli blames inefficient governance procedures, describing "confusion then existing in the Sundarbans Commissioner's office" (74).

Love Islands

By virtue of being remote, the islands appear also as spaces of refuge and of doings things differently. The two stories of horses I introduced above are instructive here. In both stories, the stately animals signal how political authority was jeopardized on theses edges of the land. The proud sahib is traumatized by the tiger killing his brother, and literally runs for his life, giving away his estate for free. While the mighty king not only loses out on his show of earthly power once the sacrificial horse gets transferred to the cave, his life also becomes enshrined in misery after losing sixty thousand sons, while his grandson lives a life of austerity. The horses are stripped of their stately quality—one left dead after the assault on its master, the other tied to a hut and grazing peacefully instead of performing royal sovereignty. Seen through the lens of the horses, customary authorities hit a wall; they seem to falter in these low-lying tracts. In this view, the image of the horse resonates with a theme used to narrate the pull of the islands. I call it "love islands," embracing the cinematic and perhaps even utopian feel of the term. If, as I have argued above, the coastal swamps enabled many to escape from suffocating density and impoverishment, promising entrepreneurial trajectories, they were also envisaged as sites where social constraints were lax. In my notion of love islands, the island boasts spaces of potentiality where settlers can indulge in illicit or unauthorized love relations, marked by a superior degree of communal harmony.

Hanging around a tea stall during a long midday break, Dilip Jana proclaimed that some 90 percent of the married couples on Sagar got engaged without their parents' consent. Once he had put out his bold claim, a grave silence set in. All present in this calm hour seemed uncomfortable yet remained in agreement. A degree of shame about not fulfilling normative expectations regarding sanctions of marriages by elders was tamed by pride in breaking norms. This came across also in more intimate conversations than public tea stalls allow for. Islanders would sometimes refer to illicit love relations as the fundament of their island homes. Rather strictly enforced practices of parental arrangement of suitable spouses, or at the very least only with parental consent, continue to characterize mainland life. Island life, however, provided openings and possibilities to transcend customs. Couples in love would flee here, I learned, looking toward good chances of establishing themselves and of blending into settler society. Put forward most often in hushed voices, this theme rubs against narratives of despair and disasters. It resonates well with accounts of frontier societies, in the Bengal and beyond (Eaton

1996), where mainstream cultures' surveillance and norms appear loosened, providing for spaces to try out new structures and test boundaries.

The construct "love islands" stretches beyond mere romantic relationships. It also covers grounds of utopian promise. Islanders emphasized frequently that the character of political relations provided for a stark contrast between mainland and island. If political discord reigned on the mainland, where violent clashes were prone to erupt, life on the island was, as I learned, considered quieter and more peaceful. Interpretations of how this came about differ. Some attributed it to the islands being remote, with political discontent and rumors traveling rather haphazardly across the gulf between mainland and island, particularly in times prior to mobile phones. Such depictions paint the islands as driven by kindness and fellowship, a certain flair that made them attractive. Conversely, the disparity between authoritarian control farther inland and laxed rules on the watery edges echoes narratives that portray the islands as an escape from the claustrophobic mainland. Both signaled island life as an alternative to drudgery, in all its challenges. Return was barely an option.

The intersections of desires to exit drudgery and the promise of the frontier appear not only in the frame of love islands. They also shine through accounts of entrepreneurial daring. Anil, for instance, claims that his father arrived on the swamps alone and found a space for himself and then another one for close relatives before he called upon home and had his wife join him in their new homestead. Similar takes on daring and achieving forefathers surfaced throughout conversations. Mobility here emerges as a form of care, a trail blazed by enterprising men, daring out into the unknown and carving out a niche for their close ones. Combining distress and entrepreneurial forms of care, such visions introduce the islands as vast spaces, ripe with opportunity, be it for making new lives or providing for relatives.

Such narrations of exploits, of entrepreneurial prowess and success, call attention to land-based visions of the good life that have since turned fundamentally untenable in the mouth of the river. They also alert us to the tensions opening up between the poles of mobility and immobility or fixedness, and the way people negotiate circuits of displacement and emplacement.

Rural Riches

Arrival was an achievement. Making it out of impoverished lands where dearth and rigid norms had become suffocating was impressive on its own, and establishing oneself in the swamp even more so. Islanders told me of the tough first years they, their parents, or grandparents went through: cutting the forest, laying embankments, waiting for the rains to wash the salt out of the former tidal swamp, and toiling to secure first harvests. All this while access to drinkable water was unreliable and dangerous creatures lurked in the woods and waters. I heard of crocodile attacks and the disappearances attributed to tigers, of harsh times marked by backbreaking labor, sweat, and scarce meals. Departing from the rather appreciative depiction of landlords as caring agents, as in Anil's iteration above, most islanders insisted that in the first couple of years hunger reigned supreme, with the absence of productive fields mitigated only by the availability of fish in the water and crab in the forests.

But fortunes eventually changed. Life histories detail the almost obscene vitality of the soil, blossoming once salinity had been wrought from it. They signal the promises of land waiting to be settled on, on what seemed to be a hidden gem of an island.

Anil waxed lyrical about the incredible fertility of the soil and the richness of nature's bounties as soon as the swamps had been inhabited, the tides tamed, and the fresh water returned. Mangoes would explode while hanging on the tree, both stones and fruit erupting with new saplings before the first fruits had even been picked. Similarly, streams were teeming with life, with fish jumping onto boats on their own accord. As a consequence of such fertility, abundance reigned supreme in homes. Milk was plenty, he emphasized alongside many others, an ingredient to every meal and not something sparingly used, if available at all, as is the case today. What are now rare and cherished varieties of fish adorned the plates at every mealtime.

Anil's voice turned remarkably warm as he told of those treasures. Others had glistening eyes too. Still, such accounts remained hidden in rather formulaic statements—rehearsed for journalists, bureaucrats, or the odd anthropologist—cohering around losses and displacement. Whenever rural riches appeared, however, they articulated visions of the good life and of ideal conditions.

Alongside excessive fertility, rural riches also manifested in the availability of open lands. Take the trope of almost instantly calving cattle, for instance. I was told frequently that in those days, cattle would roam freely—something that is off-limits today for fear of theft and in order to protect fields and gardens from grazing cows, whose indifferent browsing could easily bring conflicts to the boil (B. Roy 1994). In those days, however, cattle would disappear into the woods in the morning, I learned, and might very well return with a new calf in the evening, replicating the fertility of surrounding fruit. Cattle disappearing into the jungle, and returning well fed and with offspring, indicates how farmers profited from the forested frontier in ways that ultimately question the distinction between farmers and forest dwellers (Agrawal and Sivaramakrishnan 2000). Forest usage in present-day protected Sundarbans corroborates these overlaps (Jalais 2010b).

But it is not cattle or other fruits of the forest alone that brought the jungle's vitality to bear on plates in farmers' houses. In themselves, swamps continued to be openings and promises. Anil noted that in those days, landholdings would swell rapidly. Tiny plots encircled by swamps would grow into substantial holdings. Land would either be simply taken or bought at cheap rates, as the story goes. Fields and gardens expanded, driven by the ready availability of land that brought people's industriousness to fruition. Farmers quickly ended up cultivating twenty *bighas* or more, I was told, moving from landless to what locally counts as big landholders in just a few years. The quality of the soil combined with the quantity of land waiting to be cultivated fuel these plentiful tales of collective rags-to-riches transformation.

I like to think of this as rural riches. It seems irrelevant for my purposes to what degree narrations informed by these accounts lean toward nostalgia or even magical realism. Measuring well-being on such terms underwrites that islanders saw themselves as farmers who desired stable lands to anchor their lives on. Exploding mangoes, prolific cows, and unlimited lands all inform decidedly peasant takes on wealth. Furthermore, these narrations signal two dynamics that are critical in accounts of mobility on this coastal frontier. They articulate a momentary lapse in agrarian crises that otherwise dominate life histories. In doing so, they illuminate the kinds of circumstances and achievements islanders used to desire. More than stories of overabundance, these are stories of being in place and of having secured what Marxists call the means of production. They offer a striking contrast to the impoverishment and involuntary mobility that lurk on both sides of these episodes.

Yet they also position the good life as firmly within the reach of jungles. If the limits of arable lands had spelt trouble on the mainland, rural riches blossomed precisely by providing access to wastelands, their bounties being, as narrations show, the explosion of livestock and the virtually unlimited extension of cultivated lands. They gather their strength through the depiction of an idyllic lifestyle when, at land's end, land was not yet at its end.

Nayachar and the Ends of a Time-Tested Strategy

Eventually, in a cruel reversal, land did come to an end. *Bhangon* set in, and islanders were encroached upon by estuarine waters relentlessly pushing toward them. Prospects of settling new land on the island evaporated, and rural riches evaporated into thin air. The seams of the delta turned from a space of refuge and potentiality into a dead end.

But if the social memory of density, conflicts, suffocating norms, and normalized impoverishment rendered the mainland a hostile terrain, and if the stories circular migrants brought home from trips to cities or distant worksites invigorated notions of hostility, was there really nowhere to go? After all, the Sundarbans are famous for housing the largest mangrove forest in the world. Moreover, the archipelago features unsettled lands even beyond the limits of the nature conservation area. Exploring contestations around these lands, I show in this section that land ended also because entire landscapes became marked as off-limits through shifting state policies. I demonstrate how being trapped at land's end not only is a function of coastal erosion or of economic crisis and its ensuing dislocations, but that it also rises from governance procedures. In other words, I draw attention to the political ecology of land accretion.

As a matter of fact, islanders would have had a place to go to even as land ran out on the better-known islands around Sagar. By this I do not mean the patches of swamps still available on Sagar, which became the site of state-sponsored resettlement operations (detailed in chapter 5). I rather refer to minor islands in the vicinity deemed more or less suitable for habitation.

One of these islands, Jambudvip, made headlines. Situated to the southeast of Sagar, the island had been recognized by the state as a site of seasonal dry-fish operations until 1998. Fish workers used to rely on the otherwise unsettled lands to dry and sort fish during certain months of the year (Raychaudhuri 1980). After 1998, the government refused to issue permits, and fish workers were violently blocked from entering in 2003; the government claimed control over the island and labeling temporary camps and fishing operations as illicit encroachments (N. Das, n.d.). Fishermen organizations protested alongside urban civil society groups in Kolkata, referring to the customary status of the practice and the concomitant threat to livelihoods. In media statements, spokespersons quickly laid out what they saw to be at stake, namely that the then ruling government had earmarked Jambudvip to be home to a major tourist development project. Resort facilities on a sea-facing island very close to the Sundarbans Biosphere Reserve would, if the state had its way, tap into thriving upmarket beach and wildlife tourism.

Much to the relief of environmentalists and islanders alike, the project never materialized. But the island remains off-limits for any kind of usage. Islanders told me of regular police patrols as well as violence and penalties awaiting trespassers.

Rumors of police violence were not plucked out of thin air. The Sundarbans Biosphere Reserve is a prime example of what Brockington (2002) calls "fortress conservation." Here paramilitary forces enforce the virtual absence of humans from a seemingly untouched wilderness via patrolling and heavy penalties, mitigated only by rather rare permits. Scholars and activists document the persistently violent character of encounters between forest users and rangers involving humiliation, beatings, and extortion (Jalais 2010b; Cons 2021; Kanjilal 2000). Endemic forms of violence gave way to bloodshed during the 1979 Marichjhapi massacre (Mallick 1999; Sengupta 2011). Refugees from Bangladesh had defied state orders and autonomously begun settling the island of Marichjhapi deep within the perimeters of the national park. In January that year, security forces cordoned off the area, destroyed settlements, and killed an unknown number of people and evicted all others. The incident never was investigated. It sent a clear message about the readiness of the state to push its visions of the national park, and, more broadly, of fullforce state-sanctioned forms of land use.

The evictions of fish workers on Jambudvip were less violent, yet they carried the same message. My interlocutors knew to read the signs and shunned settling over on Jambudvip for fear of police reprisals. "It's too dangerous. Police will drive us out," I was told.

The scars of police violence were also deeply engrained into Nayachar—another, much larger and, in the eyes of my interlocutors, better-situated island. Nayachar—literally, "the new island"—is testimony to the ephemerality of land at the delta frontier. While Ghoramara, Lohachara, and Sagar, alongside many other settled islands, were subject to coastal erosion, new land simultaneously accreted into Nayachar. The island has been growing steadily for about four decades, providing ideal conditions for settlement. The land is well elevated and its position halfway between Sagar and the mainland of Medinipur promises convenient access to markets and infrastructure.

Nayachar's history also contains the fantasies of developers that never materialized, directly and indirectly. The Left Front government allocated it to be yet another Special Economy Zone, and engaged negotiations with transnational investors who planned on developing chemical complexes on the island (Bandyopadhyay 2008; Rudra 2007). The proximity to the port and chemical complexes of Haldia certainly seemed promising. Urban environmentalists, however, were alarmed for two major reasons. Firstly, they anticipated the uncontrolled release of sewage into the sensitive estuary. Secondly, given the arbitrary nature of storms and erosion in the area, the very existence of the island was anything but certain: its eventual ruination or abandonment posed threats to the sensitive ecosystem of a different magnitude.

That being said, the shelving of the plans was, arguably, less a matter of environmental concern and more one of electoral realpolitik. Ever since 2009, state governments had the debacle of Nandigram and Singur on their minds. Not far away from Nayachar, in Medinipur, farmers had seen the forceful acquisition of their land for the benefit of a proposed car factory (Patnaik 2007). Unrest ensued, which included sustained mobility and violence between police, pro-government groups, and dissidents. In due course, the Left Front coalition lost its grip on its electorate and was voted out. Large-scale industrialization projects in this corner of the state continue to be too hot to handle ever since.

In the eyes of islanders, these events continue to haunt Nayachar. My interlocutors knew that the state government was bent on keeping the island empty from regular settlers in order to have it ready for eventual development. Across the years of my fieldwork in the region, the island continued to grow. A small number of people settled there, most of the them adjacent to a now-abandoned state-run aquaculture project (Bandyopadhyay 2008, 201). But islanders on Ghoramara and Sagar took them to be exceptionally well-connected people from the mainland who obviously knew how to play the mainland bureaucracy that now had the say on the island. In other words, these lucrative development projects, despite their ultimate downfall, as well as the bureaucratic dealings of the centralized government, led to further disconnect for islanders.

Rumors also had it that during the 2007 unrest, police and hired gunmen had buried bodies of massacred opponents on the island. Living there peacefully was impossible, one of my impoverished interlocutors residing on Ghoramara told me, because of the restless ghosts continuing to haunt Nayachar.

Taken together, the threat of violence from security and the supernatural forces rendered the island a no-go area. The bulging landmass promised no relief from the troubles at land's end. The sense of being outmaneuvered was actualized all over again by the emergence of another sandbar that is now partly blocking the old ferry route between Ghora-mara and the mainland. "It will take years to develop into land to settle," a passenger told me during my last visit. "It needs to grow first." But while this was happening, security forces were policing the area to preempt unauthorized settlement. As the ferry passed the bar, we saw a few pic-nickers enjoying the open land, though it remained firmly out of reach for settlement by destitute islanders. The passenger told me that the main-land administration would settle their own people (*nijera lok*) in order to take advantage of it within the dynamics of votebank politics.

If *bhangon* meant the return of poverty, density, and ensuing forms of conflict in the homes of islanders, and if the distributed disaster began washing people along the shore all over again, the window of opportu-nities to secure agrarian futures elsewhere had closed further. Islanders ran out of options, with all jungles cleared and the remaining ones set aside for industrial projects or nature conservation. Likewise, returning to a home country known through the combined lenses of poverty, distress, and unfreedom seemed out of bounds. Thus, the question of "Where to go?" turned, I argue, into a motto to stay put, to endure and use the means available to secure a future on, and not beyond, the islands. As an expression of misery, it became a rallying cry to push for relocation. As an expression of entrapment, it underwrote attempts to make do with the means at hand. Even if they were less than ideal.

Where to Go?

More than just dusty distant pasts, the trajectories unpacked so far continue to inform vital decision-making. In debating *bhangon,* island-ers consistently referred to feeling trapped. *Kothay jabo*—where will we go? Regardless of whether we were presently concerned with individual attempts to navigate coastal encroachment or with the plight of whole islands facing uncertain futures, this phrase would unfailingly surface. It articulates a sense of being simultaneously at the mercy of larger-than-life processes and physically trapped in what feels like a dead end. Land has come to an end, with virtually no jungle remaining to be cleared, and with unsettled islands off-limits thanks to rather tight state control or visions of violence. Exorbitant prices and exploitative relations on the mainland and farther away on the Indian subcontinent only actualized vividly remembered pasts of being ousted by density, conflict, and dis-ruptions. Islanders would, at times, refer to the absence of space to live in, even speaking of not being welcome there. It is this sentiment of not wanting to go back—to now-hostile terrains left not long ago—that accentuates the experience of being trapped. This partially explains the decision by most of my interlocutors to remain on the crumpling coastal seams and to shift with the encroaching sea, instead of fully moving out.

Situating the experience of environmental displacement within what I have called "circuits of displacement and emplacement" helps to correct reductionist takes on the refugee experience. This is not to deny the specific character of getting ousted by the sea, but to enable a better understanding of how it seeps into the ordinary and the ways in which people deal with it. Troubling ourselves with situating environmental displacement on a broader spatial and temporal canvas yields critical insights for policy debates too. Using the concept of circuits of displacement and emplacement illustrates the ongoing nature of processes of tightening, loss, and involuntary forms of mobility. It alerts to nonlinear temporal horizons, to cyclical returns of acute crises as well as episodes of content. I have also shown that even if the circuits of displacement and emplacement articulate a sense of loss, there is more to bereavement. My tracing of entrepreneurial visions of selfhood and of the affective dimensions of living in frontier societies uncovers agency in an experience all too often glossed over as mute, powerless victimhood.

Chapter 5

Resettlement by Other Means

Among the steady stream of Hindu pilgrims and holi-daymakers, another type of visitors make it to these shores: those carrying cameras and notebooks, here to cover climate change realities, are perhaps less visible than the vibrant and noisy throngs of devotees. Taxis ferry journalists to what they frame as refugee colonies, and rented trawlers ship them around crumpling shores, with brokers ensuring smooth arrangements and reliable access to interviewees. Given convenient tide timings, such report visits can be completed successfully within one or two days. Alongside journalists, academic researchers also regularly end up here. People in Colony Para tell me of students from Kolkata's universities dropping by to gather data, while, seen from the tail end of academic research, a steady number of publications uses Sagar Island as a case study of displacement in the context of climate-induced disasters (see, e.g., Das and Hazra 2020; Aditya Ghosh 2017).

Alongside pilgrimage site, former jungle, and beach holiday destination, Sagar also is a point of call for a specific form of problematizing climate change. The tourism infrastructure, making arrival and accommodation relatively seamless, certainly plays a key role. More importantly, however, the island features five colonies earmarked for so-called climate refugees. Visiting it seems to enable engaging a rather unique predicament—climate change impacts, sinking islands, exile, and relocation. The socio-material form of the colony speaks in itself to the crisis narrative, while also catering to the demands of rushed visits (Harms 2018). In due course, the island has come to feature prominently in writing on sea level rise and its catastrophic consequences. It has become one of the few places dominating the debate on climate change in India and beyond. I have argued elsewhere that attending to those colonies, and the realities they inhabit, serves as a proxy for global audiences to witness climate change run its course in a particularly drastic form (Harms 2018;

Harms and Powalla 2014). Pondering coastal futures through the lens of the Sundarbans thus puts centerstage so-called refugees and the menacing question of resettlement along densely populated coasts.

The craft of anthropology knows of these messy entanglements. Ethnographers, stocked up on analytic ambitions and ethnographic passions, have long attached their work to specific locales or social groups. Particular regions appeared to feed particular conceptual debates, attracting people with specific personal leanings and interests, which, in turn, has cemented the mutual identification of places with specific problems or concepts (Descola 1998; Fardon 1990). Matters of climate change, in addition to a number of other vital debates, including genetics and pharmaceuticals, further complicate the picture. Here, anthropologists enter into such feedback loops alongside a host of other observers, such as journalists, artists, or strategists of humanitarian organizations (Pandian 2019).

In this chapter, I take this conundrum as a starting point and explore resettlement experiences ethnographically. In so doing, I address two connected claims. First, I trace the social contours of resettlement regimes, highlighting the ambivalent status of rehabilitation between bureaucratic regimes and actual relocations. I trace how relocation is achieved, bypassing concerns for climate change or environmental causes and making use of localized networks. Subsequently, I take issue with the assumption that relocations were uniform. Focusing on two colonies, I show that experiences of relocation in the Sundarbans are diverse. Remaining attentive to such diversity is integral in understanding the processes of governance, displacement, and (re)settlement in a densely populated—and undeniably strained—delta front.

The global discourse on resettlement in the context of environmental degradation is fraught. It sits among a number of vital debates in which the universalizing ambitions of conceptual work enter the rough waters of no-less-universalizing formulations on the rights of refugees or the duty of states to provide for the survival and well-being of their citizens. Not to mention the intricate matter of attributing responsibility and obligation when it comes to environmental degradation.

Those contributing to the debate distinguish between preemptive and retrospective measures (Oliver-Smith 2019; Pilkey, Pilkey-Jarvis, and Pilkey 2016), and between the resettlement of bruised populations—within their respective nations of origin—and resettlement in third territories (IDMC 2021; Dasgupta et al. 2022). They distinguish between resettlement drives surrounding mono-causal interventions, such as toxic discharge, and those that emerge from distributed interventions, such as anthropogenic sea level rise (e.g., De Sherbinin et al. 2011; Oliver-Smith 2009). Irrespective of longstanding discussions, government responses remain scant and piecemeal.

Debates on the consequences of industrial development put resettlement on the agenda of anthropological research. Scholars began tracing the politics of resettlement operations in the shadow of big dams or other spectacular projects where national development demanded the sacrifice, or so it seemed, of entire landscapes and communities. Mirroring the typical concerns of anthropology after the 1960s, studies would trace continuities and change as populations were transferred from one site to another (Colson 1971). The record remained troubling, and resettlement a trope in what Ortner (2016) dubs "dark anthropology." Resettlement operations frequently failed to recompensate, deepening the disenfranchisement of already marginalized populations, who had to make way for interventions from which other groups stood to benefit (Baviskar 2004). Debates on resettlement or relocation along rising seas rekindle many of these concerns and accentuate the limitations of approaching environmental concerns through a national lens and within limited temporal frameworks (Pilkey and Pilkey 2019).

In the last decade or so, the debate has shifted to account for the multiplicity of mobilities. Fieldwork-based accounts demonstrate that processes of decision-making are manifold and hard to pin down, and patterns of migration arguably more so. Researchers now regularly reveal migration to be not simply an outcome of failed adaptation to environmental stresses, but rather a strategy of adaptation (Black, Bennett, et al. 2011). This may involve people moving along with shifting environments, making use of whatever spaces are afforded by shifting lands or waters, or by governance regimes and infrastructures (Lahiri-Dutt and Samanta 2013). It may also involve everyday lives spanning diverse locations and lifestyles. This is the case, for instance, when people manage to thrive even under harsh circumstances by embracing mobile lives. Translocality captures this middle ground between migration and staying put (Greiner and Sakdapolrak 2013), this mobilization of counterflows of mobile people and remittances routed along webs of kinship, friendship, or "people as infrastructures" (Simone 2004). Even if such approaches are fairly new imports to the debates on adaption, ethnographic and historical accounts highlight that these do not seem novel phenomena at all. Indeed, people navigate challenging circumstances by mobilizing webbed fields, pastures, and worksites wherever we look.

Different terms pepper the debate. It is worthwhile to pause for a moment and reflect on what they imply. I read of "climate-induced relocation" (Ajibade and Siders 2021) or of "resettlement associated with climate change" (De Sherbinin et al. 2011). I mute my critique of the preoccupation with climate change (but see Hulme 2011; Dewan 2021b), which overshadows—as I show throughout this book—both the complexity of socio-ecological processes sustaining coastal erosion and of embodied ways of making sense of these on the ground. Instead, I want to highlight that these concepts prioritize forms of settling arranged for by the state that are implemented in order to govern specific groups of people. Thus, much of the debate unfolding through or along with these terms considers parameters for effective governance, state action, and inaction (Mortreux et al. 2018), or how these manifest in a kind of state-sponsored xenophobia (Hartmann 2010; 2001). Such approaches hark back to the rich scholarship on the nexus of displacement and development intervention, which frequently used resettlement schemes as a prism for considering questions of justice, societal dislocations, or affective rifts associated with state-sponsored development projects. Critical scholarship on development-induced displacement and resettlement in times of climate change remains of utmost importance, with the dangers of replicating shortcomings and injustices well documented (H. M. Mathur 2015; De Sherbinin et al. 2011; Oliver-Smith 2019).

Preemptive measures, in which people move out of reach before serious consequences unfold, play a key role here. Conceptual takes such as "planned retreat" (Koslov 2016; Pilkey, Pilkey-Jarvis, and Pilkey 2016), "planned relocation" (Mortreux et al. 2018), or "managed retreat" (O'Donnell 2022) take these concerns explicitly to the coast, and into terrains marked by coastal erosion. All emphasize dedicated state welfare programs, or the lack thereof. This leaves a great number of relocation practices, and their associated experiences, out of view, especially in the Global South, where relevant governance regimes many not be in place and people regularly relocate haphazardly (Ajibade and Siders 2021; Ajibade, Sullivan, and Haeffner 2020). Numerous studies embrace planning as an activity of personal or household-level decision-making in the context of navigating deteriorating environmental or economic conditions (Ajibade and Siders 2021). However, the propensity to concentrate on state interventions tends to considerably eclipse any unaccounted for or unsanctioned forms of relocation. If the field of resettlement or relocation could be divided along a temporal arc—dividing punctual events from chronical effects, or preemptive measures from post-hoc interventions—these temporalities would, in my reading, intersect and give way to patchy scenarios where overdue resettlement drives are shelved and people excluded from drives feel trapped, chronically living in danger. In this chapter, I turn to a complementary figuration, and show how bureaucrats and victimized populations roll out resettlement drives on their own terms, and how they make use of normalized land distribution to answer to the pressing demands of fellow islanders washed into misery by *bhangon*.

Scholarly and societal debates on retreat along sinking shores seem to prioritize settlement regimes explicitly based upon "climate change" or "sea level rise," and "displacement" or "relocation." This is all very important in a context where ever-rising numbers of displaced people face an absence of binding regulations or reliable procedures (Neumann et al. 2015; Kulp and Strauss 2019). However, I suggest that considering how resettlement occurs—through makeshift relocation or retreatment schemes—illuminates unchartered aspects of the refugee experience. Moreover, it signals that significant relocation in the wake of environmental displacement might not yet have been officially documented as such. Finally, attending to the ways in which resettlement after environmental displacement overlaps with normalized settlement schemes allows us to add historical strata, as well as retrospective accounts, to a pressing problem. In a curious twist, my chapter thus speaks to the concerns of debates on planned relocation or managed retreat from a place of abandonment and improvised forms of resettlement. In their messiness, these accounts help us to reconsider the experiences of rehabilitation. Alongside these theoretical ambitions, my account highlights the power of localized networks and the shared sense of belonging involved in successful settlement operations. It thus aims to reorient the debate toward local and regional frameworks.

Unacknowledged Forms of Rehabilitation

Resettlement was an anomaly. Retrospective news coverage or scholarly work attests to that. It was an anomaly not because the process it seems to answer to—environmental displacement—was unique or novel, but because this mode of rehabilitation was seldomly carried out. India's record on resettling displaced people is poor by any given standards. The question needs to be raised of how people secured relocation, and even more so of how they did that when displaced not by a development intervention but by distributed disaster. The imagery of eroding homes and submerging islands, carrying a particular force, might play a role here. The persistence of islanders, voicing their suffering in the form of written requests and putting pressure on politicians, probably also has its part to play (although my account clearly shows that such forms alone are nowhere enough to succeed). Importantly, I argue, this anomaly needs to be placed within a specific political figuration. By this I mean the prioritizing of displaced islanders within a small spatial and administrative unit of the island group as beneficiaries—a mode that was fueled by quests for allegiance in electoral politics, by the sentiment of common pasts shared at this frontier, and, finally, by the ambitions of emerging leaders who use land distribution to their advantage.

Disagreements exist on the architecture of the resettlement operations. Pramanjan Mandal, for instance, who had served as Sagar Block's MLA[1] on a CPI(M) ticket, claimed in an interview that resettlement operations where the outcome of a "colonization scheme" that formed a segment of Indira Awas Yojana (IAY), the famous long-standing program to provide shelter to the poor. Most politicians and bureaucrats to whom I spoke, however, emphasized that resettlement operations were organized within the rules and regulations of land reform, and that the instruments of the IAY had been used complementarily in some cases. Bureaucratic files, and the procedures through which they wielded power, corroborated the latter claim. All islanders received land titles through the local Gram Panchayat office, with the files connected to these titles speaking the language of land reform.

A pet child of Bengal's left-wing government during the advent of its reign, land reform provided powerful imagery and a set of governance procedures that bureaucrats made use of. As a bureaucratic practice, it rendered the distribution of land to landless citizens desirable and doable, connecting it to a project of social justice. A textbook case of leftist attempts at decolonialization, land reform sought to roll out justice and mitigate poverty by way of distributing land controlled by large landholders, or unused lands, to the deserving poor. Ambitions were universal, and, in theory, all the landless eligible. Studies of land reform across the state show that while substantial lands were distributed, the reform quickly became embroiled in the nitty-gritty of party politics. In fact, land and papers turned out to be a means of gaining votes in fraught elections. Land reform, in other words, was seriously constrained and contained

---

1. MLA stands for member of the Legislative Assembly, voted to represent a district to the state government.

within local arrangements of power (Lieten 1996; Mallick 1993). Alongside these critiques, without attempting to gloss over the injustice committed in the actual rollout of land distribution, I maintain that these constraints were also productive and allowed for the distribution of land to environmentally displaced people in a process I call "unacknowledged relocation."

Central to these processes were lists compiled by the staff of Ghoramara's Gram Panchayat office. Staff collated the names of people eligible for sections of land. They did not detail circumstance of displacement or, as a matter of fact, whether the listed individuals had been displaced at all. Instead, they listed them as destitute people, landless citizens hailing from the vicinity who were deserving of state welfare in the form of land allotments. By the time the resettlement drives took shape between Ghoramara and Sagar, however, only islanders ousted by the estuary made it onto the list. Over the years, such lists propagated. They were not only tools of state welfare, nor were they mere instruments of belonging to the polity. Below all that, they were, fundamentally, means of making the regimes of land reform do a specific job, that of rehabilitation after environmental displacement. For an extended moment in time, land reform attended to the needs of certain islanders, but ultimately to the detriment, I might add, of all other landless subjects. To be clear, this was nowhere near corruption, but a process that accommodated one type of landless people at the expense of all others.

If the very compiling of lists qualifies as an instance where people mobilized bureaucratic tools to serve their demands and aspirations, these graphic artefacts gained worldmaking powers through a set of inscriptions and further modifications. As is typical of bureaucratic governance in India, the lists needed to pass through a series of offices and stages of assessment, copying, and authorization. They gained power through a set of inscriptions (Hull 2012) as officials added signatures and stamps onto files to render them valid and effective. From here, they became devices of "future histories" (Ballestero 2019), of engraining potentialities and prefiguring futures.

In order to stay within the contours of land reform, to make the lists serve their function, it was necessary to ensure that *unacknowledged resettlements*—in which environmentally displaced people found divergent means to continue their rural lives—remained hidden. But if—according to governance regimes—those walking out of Gram Panchayat offices with fresh land titles in their hands were simply the rural destitute, many islanders worked against such framings, instead claiming a degree of visibility for the specifics of their resettlement. In ordinary conversations, they would demarcate individual colonies as ones set aside for the resettlement of victims of coastal erosion. Such takes foster the notion of resettlement as a spectacle, while rendering beneficiaries as a somewhat distinguished category of the poor. It is likely that this differentiation was ingrained in islanders' minds from the first days of resettlement. It served as a demarcation between them and their new neighbors, setting them on another plane of destitution than those who are often referred to as ordinary street people (*rastar lok*). Yet, the stream of media persons, researchers, and activists trickling in from the wider world, to get a glimpse of this rather unique form of displacement, continuously reinscribes such a distinction.

Authorization between Spectacle and Territorial Units

Colony Para was replete with stories of help received from the highest offices. Resettlement, in this view, was no smooth or mundane affair, not something arranged for through the lower capillaries of power. There was reference to the Indian prime minister paying a visit to the area, hovering over the island in a helicopter, from where he witnessed the miserable state of the islands of Ghoramara and Lohachara and the plight of its inhabitants. What he saw, I was told, made him push for quick and comprehensive relocation to be rolled out by his subordinates. Others were certain that the resettlement scheme was sanctioned in India's capital. Sheikh Mumtaj, for instance, insisted that all files were sent to Delhi and authorized directly by ministries of the central government.

Skimming through the paperwork, including research and media accounts, I failed to find further evidence for these claims. In fact, as I will argue, relocation seemed to prove effective precisely due to being a local affair, contained within the limits and the bureaucratic routines of a small administrative unit—that is, the Block. However, from islanders' accounts, it seemed as if allies from above—literally in the sky or in the highest echelons of power—put the gears of local bureaucracies into motion. These claims sit well among a shared consensus on the limited care provided by the local states and the many frustrations citizens in rural Bengal experience in their everyday encounters with the state. To some, the central government appears as a corrective, especially since the rift between a more center-right central government and a more center-left state government used to be an acute one before the 2011 watershed elections. More fundamentally, I suggest, the claims above lend additional weight to the plight of the islanders, once again rescuing their fate from that of ordinary landless poor citizens populating the state in great numbers. Environmental displacement—as a spectacle and a fate its victims are not responsible for—is rendered distinct, apart and somewhat above and beyond ordinary forms of destitution in a persistent rural crisis.

Irrespective of such spectacular feats, lists traveled widely. Authorized copies traveled to the Block headquarters, the Block Development Authority (BDA), and from there on to Kolkata. And it is necessary to consider the technicalities of the administrative setup in more detail. Sagar Island, plus Ghoramara and the now submerged Lohachara, comprises the Sagar Block. The BDA, for most purposes simply called the Block, serves as the lowest rung of the state government. Blocks are both administrative units and bureaucratic machines, tasked with rolling out rural development, helmed by members of the prestigious IAS (the Indian Administrative Services), sent by the state government. Sagar Block comprises nine Gram Panchayats, whose staff are not sent by the state government, nor do they answer to them, but are elected in distinct elections. Both are vehicles of party politics, yet the arithmetic plays out differently. With the Block being a function of the outcome of statewide elections, the Gram Panchayats mirror the results only within the territory of the Gram Panchayat. Formally independent, they are bound together through development visions and state welfare programs. They are crucial avenues for territorialized belonging as well as the everyday politics of inclusion and exclusion in areas pervaded by abandonment.

In Calcutta, the Ministry of Land Reform approved claims for the distribution of "unused" land, sending authorized lists of beneficiaries back to the Block headquarters. In all of this, the Block administration served not only as a key focal point. The rather narrow territorial and administrative frame of the Block proved crucial, I suggest, for the establishment of colonies and the successful rollout of relocation. Officers of the Block coordinated the identification of suitable lands, unclaimed and vested in the state, needed for the relocation efforts. It facilitated exchanges that bounced between outpost offices and ministry buildings.

On the surface, the state issued land titles to victims of coastal erosion. Islanders walked into Gram Panchayat offices, and by virtue of having their names on relevant lists, received papers that allotted plots to them. Viewed differently, the islets of Ghoramara and Lohachara were slowly being submerged, and as a consequence of this, the population of one of Sagar's nine Gram Panchayats, comprising these two islets, was thinning and the administrative unit itself dissolving. Meanwhile, other

Gram Panchayats in the vicinity, in the center and the south of Sagar Island, were growing in numbers through a process of relocation. Considering the next administrative layer, people moved from one end of what is known as Sagar Development Block to the other. To frame this as a mere story of land lost at one end of the Block evened out by availability of land at the other end would miss an important point. The Block served as a frame small enough to see the relocation from one of its ends right through to the other. It enabled the rise of new political leaders who, in turn, managed to make the most of the close circuits of these two entangled administrative orders, a point to which I will return below when I discuss the fate of individual colonies in more detail.

Resettlement was not only an anomaly—what it achieved was also subject to substantial changes. The size of land parcels was reduced repeatedly; the scope of distributing land gradually diminished; and the process of limiting resettlement to environmentally displaced islanders was getting more and more porous over the years. Across all these changes, however, procedures remained strict for all practical purposes as to who was eligible for receiving land and titles. The rule is simple. Only one person from any given affected household could secure a plot in one of the colonies.

Greatly restricting the number of benefactors from every household is a thorny issue. It underscores acute land shortage. More importantly here, it flags the tricky questions of what a household is, where it ends, who speaks for it, and who is excluded. Such outcomes lead to contention around questions of control of land and homes, empowering individual figures as legal right holders.

Studies of governance practices in India demonstrate how normative ideals are built into them. A number of instruments and schemes if not create then at least actively cement families as units. Individual members delegate themselves to be the head of certified groups to reap any meager welfare benefits provided by the state. In some cases, what counts as a family mobilizes the image of nuclear families, consisting of parents, their offspring, and, perhaps, grandparents (Rao 2018). Attributing welfare packages to such units, and assuming them to be shared among units, inscribes predetermined visions of populations.

The act of distributing land titles to only one member of a family, for the sake of the family, produces similar outcomes. Even if this mode of distribution does not enforce nuclear families (but see below), it still enforces families as units of care. Feminist critiques, in particular, show that the family is often not the haven of care and rest it is often pitched as, but a frame of exploitation, power, and control. Distributing welfare to family units may only aggravate such power relations, giving people the control of deciding who is to earn where and how wages are to be used by the family. Or, as in the case of land distribution, who stands to become the person officially holding the land.

Juddhistir Jana, one of the first settlers in Colony Para and now an old man, had the title issued in his name. This was somewhat of an exception. Most in his generation, who by then already had grown-up sons, had titles issued in the name of their firstborn sons. Having the younger generation's names on the papers would save them the troubles of posthumous title changes. What appears as an early inheritance actually was a nod to the notorious difficulties people anticipated in engaging with bureaucracy, which could be avoided by passing ownership onward earlier in life. It exemplifies the maneuvering of papers and plots, and the way welfare distribution—and the imaginations underpinning it—rekindles inner household tensions.

By the same token, land distribution due to scarcity revised customary inheritance rules, which understood firstborn sons to be the sole inheritors, at the expense of all other children. Women were rarely beneficiaries or title holders; in all of Colony Para, there was only one such woman. And only two brothers managed to secure a title each. The procedures favoring one person as a sole benefactor for a whole family thus produced a range of tensions. Women were disfavored, not by the regulation but because patriarchal norms at play favored men. Later-born sons were also disfavored in the uneasy alignment of legal categories and inheritance laws.

Now, three decades after unacknowledged resettlement, the outcome is mixed. Nuclear families cohering around male heads have been legally sanctioned as right holders, while affective relations within larger family groups continue to undermine this picture when relatives without claims to titles manage to secure a seat at the table. Unsurprisingly, at present, Colony Para is crowded, troubled not only by ever-decreasing farmlands but also by domestic tensions, as families of later-born sons stake claims to the land. In many cases, their demands for a plot to build a house for themselves are successful.

Despite being consistently challenged, regulations and practices reinforced the individualized ownership of land and cemented gendered hierarchies. Lists were involved in politics both ways—they produced a politics of inclusion by focusing on a universal regime in favor of one particular type of landlessness (victims of coastal erosion), and their implementation reinscribed norms built into the procedures of state welfare. That is to say, seemingly neutral lists carried a politics that arose at the intersection of official regulations and lived bureaucratic routines enacted by situated actors.

If lists and papers performed a politics, the settlement of swamps in designated colonies inserted further politics of resource access. To these I will now turn.

A Death Foretold

Anil Patro's fairly distant neighbor and longtime acquaintance Jotthisdir Maiti still had a choice in where to settle. It was, of course, a decision full of grave consequences. Contrary to depictions of refugees as hapless victims, Jotthisdir considered carefully what would be best for his family before making the call. When it was his turn to receive plots and papers in 1986, land was distributed in two colonies. One in Gangasagar, where he eventually settled, and one in Bankimnagar, some thirty miles away. Together with others, he went to see the respective swamps to get a rough idea of the futures awaiting them. On the surface, the areas seemed similar. Both were mangrove patches, bordering open estuarine waters, met by tides through inlet canals, and no human habitation. Both were patches, that is, until then bypassed by development. On closer inspection, however, they differed. The major difference, for Jotthisdir, was the safeguarding of the land. Bankimnagar was much less protected from the reach of the sea, unlike Gangasagar, which was partly secluded behind a dune beach and framed by what locally would account for rather stable lands, which already boasted a village on one side and a temple on the other.

Instead, the swamp in Bankimnagar was flanked by the wide river, the Muriganga, flowing along Sagar's eastern side, itself so expansive that the opposite bank was barely visible. It felt much like Lohachara, the lowland at the mercy of these vast estuarine waters. In other words, it lay exposed to the influx of tidal waters. Establishing and fortifying the edges against the vast water masses flowing along was an uphill task, given the technical means at hand. The typical mud embankment, enforced by wooden and bamboo frames, could not quite provide protection for homes and fields. Jotthisdir, of course, was not the only one who quickly recognized this. In fact, the government had granted substantially more land in Bankimnagar than in Gangasagar because of the inferior quality of the land caused by continuous saltwater incursions, which further hints at the unlikeliness of it being thoroughly protected. The site's vulnerability to water encroachment, combined with the state's likely lack of interest in providing proper protection, made Jotthisdir turn away. To him, it was a safer bet to pass on the four *bighas* offered in Bankimnagar and accept land in Gangasagar, even if it was less than half the size.

Less visible in the beginning were the strained relations between newcomers in Bankimnagar and established villagers. The older parts had long been settled; in fact, Bankimnagar features among the oldest continued settlements of the island (Maiti 2008, 49–50), with its inhabitants heavily relying on the swamp readied for clearance. The forest provided shelter from storms und surges. The bone of contention, however, was the canal passing through the swamp. They used to land their boats here—boats that took some of them to the markets and towns on the mainland and many others to the fishing grounds across the estuary. More than a convenience, the canal was a bottleneck and a lifeline. Once they got wind of the relocation plans, villagers quickly realized that the new settlement meant an end to the canal. Blocking the entrance kept the tides out, which in turn kept the water fresh and free of salt. What is more, if blocked, the expanse of the canal could itself be turned into land. As villagers knew, this would not only increase the number of plots available; it also resonated, I suggest, with aesthetic desires informing high modernist forms of governance bent on replacing crooked, unruly canals with uniform expanses of neat plots.

As the demise of the canal could not be prevented, relations with arriving settlers turned sour. The tensions around the canal even materialized into a deadly mishap, entangling canal, swamp, villagers, settlers, and ghosts in uncanny ways. It is worth engaging with narrations of this incident in some detail as they offer insight into the struggles for settlement as well as the overarching legitimacy of the colony.

Today, people in the Bankimnagar Colony look back at their early days here as a time rife with hostility and hardship, tarnished by memories of the denial of drinking water. A patch of mangroves on swampy ground soaked by tides, Bankimnagar initially lacked drinking water. It took years until hand pumps were built. In an area like the Sundarbans, where upper ground layers tend to be salty and freshwater aquifers are present only at depths of around two hundred to three hundred feet, laying pumps requires substantial means that poor colonists can hardly afford, unless they are provided by the state. Back then, public pumps were few and far between. Ponds provided water for most purposes. People drank it and used it for cooking and bathing. However, ponds were typically private property. During those tense times, villagers from the long-established Bankimnagar denied the new arrivals access to the ponds, thus suspending customary norms of sharing water and hospitality. While the initial attempts to block access to public pumps proved impossible, ponds remained off-limits, and water extremely scarce.

Relations came to a boil when the canal was to be dried. Engineers tasked with developing the swamp into a colony had decided that the usual mud embankment was not sufficient to effectively block the entrance of a canal that wide and deep. They decided to sink long wooden poles into the riverbed, adding layers of mud around and on top of them. After the operation was sanctioned, Ganesh Mali, a local leader, supplied and stored the planks needed. The night before the works were to commence, he was plagued by terrible nightmares. Ghosts threatened him, demanding that he find ways to prevent the blocking of the canal. "How can we live here if we do not embank the area?" he asked the ghosts, my interlocutors told me. To which the ghosts only added another threat: "No, whoever piles up the mud will die!"

Ganesh did not know what to do. He shared the dream with his fellow settlers. Many agreed that the threat was real and that they'd better heed the warning. But they had reached an impasse: they required the land to be able to settle and flourish, and this in turn required the blocking of the canal urgently. Plus, it had been decided by the government. Who were they to stop it? It was agreed that they would start the work as planned. Anxiously, the laborers assembled and began their task. However, as soon as they had anchored the first pole into the mouth of the canal, the foreman received news from home that his son had drowned in their pond. All work was stopped immediately and recommenced only after an elaborate ritual, a puja provided by Kartik Misra, one of the very few Brahmins who had relocated from Ghoramara to a resettlement colony. While none of my interlocutors remembered the particularities of the ritual, they considered it instrumental in pacifying the ghosts. Following the puja, work recommenced, with no further accidents occurring. Soon embankments sealed off the whole terrain, and the swamp gave way to roads, homesteads, and, eventually, fields.

Ghosts are deeply embedded within Bengali folklore, their stories a popular genre of fiction and legend. Such tales of apparitions stretch far beyond these fluctuating coastlines. Sometimes they are considered creatures of the wind, while in other iterations they appear as very much localized entities. So-called wastelands are known to harbor ghosts hovering in trees and themselves becoming part of the places they inhabit

(Blanchet 1984, 54). In her study of Anglo-Indians dwelling in the ruins of the Indian Railway ecosystem, Laura Bear (2007) shows how ghosts haunting the railway barracks are intimately tied to assertions of self, identity, and a sense of continuity. Tangible presences in modernist ruins, ghosts appear neither as burden nor threat, but rather signifying roots to this marginalized community that has its history largely denied. In the story of resistant ghosts and a death foretold, as shared with me in Bankimnagar Colony, ghosts take on a similar role. Here too they appear as localized agents, literally engrained in place and embodying continuity. When they spring into action, it is to upend transformations that threaten their existence.

Not directly tied to the villagers of old Bankimnagar by kinship or other ties, ghosts nevertheless joined the villagers in their opposition to settlement operations. Both are hostile, threatening survival indirectly by denying water or causing "accidents." Relocated islanders have had to wrestle their future from them, toiling thirstily under numinous threats in order to anchor themselves. Simultaneously, they had to displace other entities. At this juncture, relocation intersects with displacement. Not only, then, is one man's loss another man's gain, but one man's settlement can literally mean another's disrupted access to key resources, such as a protective forest or a canal out to the sea.

The tales of threat, drowning, and then pacification, combined with those of thirst and despair, eventually enabled the crafting of a place and its population. Fused together, everyday struggles and harrowing tragedies ultimately inscribed this relocated population into place, and it established itself as a unit. To be sure, tales of the supernatural can offer only one perspective in the layered histories I have described already. Yet, and this is the point I want to make, it is a past sustained by opposition to haunting presences and conflict around the all-too-often hot issue of access to natural resources. Through this past, belonging was sooner or later crafted by the settlers in opposition to the hostile forces surrounding them.

Taken Along in Development
Visiting the swamp that would soon be Gangasagar Colony, Juddhistir Jana quickly realized the advantages of that area. For one, he told me, it was tucked behind a dune. Even if tidal waters entered the swamp via a canal, the area seemed to be better protected than Bankimnagar. The canal's mouth, opening between the dune and the neighboring village of Dabhlat, was considerably less vulnerable to currents and waves

than the planned embankments at Bankimnagar that were expected to shield the whole water-facing side of the colony directly from the estuary. Being a salty place (*nona jayga*), soaked by the tides, surrounded by troublesome wilderness (*jangal*), Gangasagar offered a sense of stability and fixity that Bankimnagar did not. It was farther removed from the estuary, he said, speaking to the desire to live as far away as possible from the relentless blue mass.

Geographical matters aside, life in Gangasagar also promised Juddhistir access to reliable sources of income. He was not alone in that. Some forty years ago, when settlers began relocating here, pilgrimage traffic was considerably less established than it is now. Still, the yearly winter festival already attracted large crowds from all across Eastern, Northern, and Western India, with a steady stream of devotees making it to these shores throughout the year. Juddhistir spotted, as many others did, the potential for this flock of visitors—moving to their own rhythms and demands—to fill critical gaps in household incomes.

And he was right. If the first years meant hardship, pilgrimage traffic brought some relief while Juddhistir and his neighbors were draining the swamp and building yet another home, this time in what was to become the oldest, most interior parts of Colony Para (see figure 5.1).

Others were more vocal about hardships endured as they prepared fields for the first time and waited for initial meager crops to emerge from

Fig 5.1. Homes and harvests in today's Colony Para, Sagar Island. Photo by author.

still overly salty soils. Perhaps jaded from the initial troubles of establishing themselves in the colony, newcomers were braced for extended lean periods. Ganesh, like many others, did not elaborate on the back-breaking labor but highlighted the absence of protection and nourishment. He spoke of days in the jungle and long nights under the stars with no proper house to retreat to, of long walks to find fresh water to wash off sweat and dirt from aching bodies, and of grueling hunger and aching bellies. To survive, they would fish or collect crabs and whatever fruits of the forest they could lay their hands on. When hunting and foraging came to nothing, they would stick to the beach and search for the coins or jewelry that pilgrims gave to the waters in order to secure boons or to return a favor from Ganga. Raking the sand with forks or simply watching out amid pilgrims were some of the more humiliating tasks, a last resort on par with begging, the personification of despair.

Whenever possible, Colony Para settlers would transact in services, offering rickshaw rides or hawking trinkets or snacks to pilgrims. Settling within walking distance to the temple made this easily possible. Svagat, the self-proclaimed architect of Colony Para, still ran a roadside restaurant during my early research years. Others who started out tapping into the pilgrimage traffic by plying cycle rikshaws now have their sons steering electric rikshaws or working on buses with the same clientele.

Investment into the development of Gangasagar as a major tourist attraction (for believers or others) intensified considerably after the Communists saw electoral defeat in 2011. Prior governments, including the Left Front, had also devoted substantial funds to infrastructure and utilities to cater to the large number of pilgrims flocking to the island from across India. Providing labor to public works or private enterprise before, during, and after the festival had been a survival strategy and an opportunity for settlers in Colony Para from the very beginning. Yet there is a dearth of reliable data on government work expenditure or on the funds traders setting up their shops on the festival ground spent on local labor.

For much of its history, the temple of Gangasagar was sandwiched between muddy jungles and the shore. To colonial visitors, it was forlorn, a remote jungle at the continent's edge harboring less-than-humane practices. To pilgrims, however, it stood out as a site for fulfilling promises made to the divine, for submerging oneself in purifying waters at the outer contours of the sacred homeland of the Hindus. Across these views, Gangasagar was a site of risks. Administrators emphasized the endangerment of babies and souls in this abyss of worship. On the other hand, pilgrim accounts, across manuals, films, and oral histories, underlined the risks of traveling to this remote outpost of the Hindu motherland. Manuals spoke of tigers lurking in the jungles. Films depicted shipwrecks in the estuary's treacherous currents (D. Bose 1959). While the colonial government did little to hedge risks—concentrating its efforts at this coastal edge on the spreading of estates, the establishment of port infrastructures, and the outlawing of infanticide—the postcolonial state slowly moved on to a different course of action.

A wide road was built. Reliable ferry services were established, reinforced by extra vessels brought from Kolkata for the duration of the festival. Electricity was provided on the island much earlier than in other parts of the Sundarbans, even if initially only along the sole road. The festival itself required massive labor input to get up and running. Toilets needed to be dug, cables laid, temporary structures assembled and dismantled.

These tasks provided income for settlers and continue to play a key role within most chronically cash-strapped households of today's Colony Para. They also embody a politics of development, and its relevance for the trajectory of resettlement in Colony Para.

I am well aware that I am giving precedence to survivor stories here. They echo the stories of achievement and entrepreneurship through which people remember how their ancestors initially began settling swampy islands. Complementing attention to the exclusions engendered by both the politics of development and the way survivors remember it, I want to foreshadow how the distribution of welfare undergirded the different experiences of resettlement between the colonies in Bankimnagar and Gangasagar.

To account for such welfare distribution, I turn to a trade union office and a water pump. This is not arbitrary. Both sites continue to be critical for the sustenance of everyday life in Colony Para. Tending to them—in a discussion of the experience of resettlement policies amid rising seas—sheds new light on two figurations that have sparked a lot of debate in theorizing politics in India and beyond (see, e.g., Anand 2017; Chatterjee 2011; A. Roy 2003).

As was the case in Bankimnagar, fresh arrivals to Colony Para also depended on water resources from their new neighbors. But in contrast to that swamp plagued by tensions around canals and ghosts, new settlers in Colony Para could rely on the pumps and ponds dug in the pilgrimage complex, thereby circumventing troubles with established villagers. And similarly to Bankimnagar, the prospect of draining the swamp didn't go unchallenged. Prior to new settlement, the swamp had been leased out as a fishery to villagers from Dabhlat. To make room for the colony, the government canceled leases. This caused fury among former leaseholders, who saw key economic practices upended for the sake of more people flooding the coast. But tensions led nowhere. The state ignored former leaseholders, who realized they needed to look elsewhere to extract fish, crabs, or wood. Likewise, the initial settlers were lucky enough to have geography on their side. Dabhlat sits on the other side of the canal. That meant that frustrated villagers were not in a position to block or limit access to any infrastructure available around the pilgrimage complex to settlers taking over the still-vacant lands between their village and Gangasagar proper.

Even though settlers benefitted from the—by local standards—lavish infrastructure provision in the pilgrimage site, not all was well. Uneven distributions took hold. Water pipes supplanting hand pumps were laid only in 2011 and never put into service. Similarly, electricity was only provided to Colony Para from 2016 onward. Settlers, therefore, had to negotiate the gap between getting provisions from the pilgrimage center and its immediate environs for decades. Yet—and this is a point I want to emphasize—the core gathering of basic provisions continues to emanate from the pilgrimage center. Since settlers were able to tap into the resources made available in Gangasagar, tensions with villagers from Dabhlat were of little consequence.

Welfare provision through the pilgrimage center was not limited to water. Nor was the presence of pumps, ponds, or roads the only tangible trace of development important to the resettled islanders. Colony Para was from the very beginning oriented toward Gangasagar—an extension of it, as its other name, Gangasagar Colony, indicates. By local standards, it was adjudged to be a spectacular development. This meant that former leaseholders were sidelined, their discontent buried by the development of the sacred center.

Arguably, the development of this shore was, for a long time, not a priority. But this slowly changed over the decades, and Sagar Block gradually emerged as the most developed administrative unit of the Sundarbans, and the southern sea beach housing the main temple as the clear focus of attention. In recent years, any bias toward Hindu ritual practices has given way to accepting the combination of pilgrimage and secular seashore tourism. Scholars across the social and natural sciences have demonstrated the profound impact of ritual or secular tourism traffic on the socio-ecological landscape around Sagar's holy beach, emphasizing local benefits, in the form of tourist income, as well as dislocation, in the form of excess solid waste or bacteria (Dasgupta, Mondal, and Basu 2006).

To gauge how these forms of development intersected with resettlement and gave a specific texture to the experience of rehabilitation, I turn to the trade union office. For decades, local trade unions have been instruments of the development that accompanies this steady flow of pilgrims. Instantiating the politics of party allegiance West Bengal is renowned for, party-affiliated trade unions would exclusively organize labor required for all kinds of festival-related tasks. Every year, labor arranged by the trade unions remakes paths lining the vast fair grounds, sinks temporary toilets, and mends electricity lines in the weeks leading up to the fair. After the fair, bamboo shelters need to recycled, and lighting and surveillance devices disassembled and stored. Thousands of shifts are distributed in the cold weeks from December to February, with the peak of the festivities taking place on January 14.

As the festival approaches, nondescript trade union offices, placed in the middle of the festival ground, spring to life. Crowds gather around the buildings, built from bamboo and tiles, colorfully adorned with flags and slogans advertising the political party they are affiliated with. Gleaning the rising and sinking fortunes of the respective political parties is a fairly rudimentary endeavor. Bodies would assemble tighter around the offices of the most influential party. The arithmetic is simple and works both ways: the party in power pours funds into "their" trade union, binding people into loyalty and enforcing votes; but when the favors shift, as they did in 2011, the spending power moves next door, to another trade union. So, while a slew of different state departments identifies tasks and allots funds, party-affiliated trade unions act as gatekeepers.

Studies of the politics of West Bengal and beyond have explored such trade unions as the entrenchment of political power in ways that circumscribe or complement ideals built into notions of electoral politics. They join the ranks of institutions that serve as means of tapping into the circuits of legitimate politics for people marginalized in society (Chatterjee 2006). At the same time, they are means of assembling crowds and count for electoral politics, in well-documented dynamics of vote-bank politics.

Most importantly for my discussion of novelty, however, trade unions also serve as machines for cultivating a sense of localized belonging. The hectic days immediately before and during the fair belied this idea. At these times, the crowds assembling in front of trade union offices featured people both from the immediate communities and from farther afield. Last minute tasks required the trade unions to open up to all kinds of people. Furthermore, the rule seems to be generally suspended when it comes to tasks such as cleaning toilets, a severely degrading task taken up only by groups at the lowest echelons of the caste system.

Quieter days, however, in the sustained buildup to the festival as well as its long aftermath, when there were fewer immediate tasks, painted a different picture. The more tolerable jobs were carried out by work gangs hailing exclusively from around Gangasagar. In fact, being less pressing, these tasks could then again be spread out, becoming a reliable source of income across a longer period of time. Workers from elsewhere rarely made it onto these lists.

"The work is only for local people [*desher lok*]," said Ganesh, echoing many others. Making it onto the list of beneficiaries of shifts was therefore a matter not only of political allegiance but also one of territorial belonging. In fact, these tend to inform one another (Chandra, Heierstad, and Nielsen 2015). In colloquial Bengali, the notion of *desh* is extremely elastic. Depending on context, it may refer to the national territory, to the federal state of West Bengal, to the coastal tracts, to Sagar Island, to the village, or to the immediate neighborhood. When discussing work for the festival and being entitled to shifts, *desh* implied only people from Gangasagar and the immediate surroundings. *Desher lok*, in other words, was used to refer to people calling the southern-central areas of the island their home, to people from *here* in a very narrow sense.

Distinctively exclusive, vernacular usage in the context of entitlement still offers an ambivalent take on what it means to belong to the proverbial "sons of the soil." If almost all my interlocutors considered their ancestral home to be the mainland across the estuary, despite many being born in Ghoramara or Lohachara, they were themselves relative newcomers to Gangasagar. Their status as residents of Colony Para makes that amply clear. The notion of *desher lok* involves banking on territorial belonging as much as it implies the efforts to create such. Therefore, I see the self-ascription of *desher lok* as signaling attempts to contain those brought along by development and to protect access to funds or indirect benefits by harnessing notions of ancestry. This not only articulates the benefits of belonging to the new terrain, it also echoes the politics of being included on the list of beneficiaries of poverty-alleviation programs at play in embankment-maintenance schemes.

Alongside income from the festival, benefits also include recurring infrastructural development initiatives. Such initiatives were still focused on the sacred complex, and a number of them turned out to be nonstarters. As noted, early on in my fieldwork, workers laid water provisioning pipes, for instance, which were meant to connect Colony Para to an emerging water distribution network, making use of Gangasagar's pumping stations. Residents managed to secure shifts in these efforts. Yet, the pipe system was never activated, the plastic pipes left rotting in the ground. In the end, the overall plan of providing drinking water in a networked fashion failed. Throughout Gangasagar, drinking water continues to be arranged through wells and pumps, from which it is carried to individual homes. I never quite figured out where the problem lay. The very fact, however, that Colony Para was included in the efforts alludes to what I call "sharing the new." When workers ripped open land at the side of the roads in order to lay pipes, these efforts stopped at the outer limit of Colony Para, at the edge of the *khal*. Settlements farther away in this direction, such as Dabhlat, were not included. It lays bare, once again, the close affiliation between the residents of Colony Para and the pilgrimage center. The construction embodied the promise of connection to a water network (Gardner 2012), while simultaneously signaling a connection to the complex and rather rich development that comes with such projects, if only by way of securing employment and bringing state investments into the colony.

As people in Bankimnagar found themselves being cut off from key infrastructures, settlers in Gangasagar benefited from access to an emergent center and its institutional backing. In a region generally lacking development or sustained government investment, access to jobs, to benefits, and to the politicians implementing such investments meant a key resource and an emblem of power. In their own ways, mundane pumps and trade union offices embody a specific politics of welfare distribution, whereby resettlement in Colony Para—organized within the double framework of land redistribution and the colonization of jungles—came to be intertwined with the overall development of Gangasagar. The sacred complex not only figured as a driver and object of local infrastructure development, it also served as a node for ongoing infrastructural development that subtly shaped settlement. Few people saw these benefits coming. They turned out to be critical, for generating much-needed cash and for the general experience of settlement.

This is not to suggest that (infrastructural) development is the sole foundation of more or less peaceful resettlements, but that islanders resettling in Gangasagar were fortunate to have been, and continue to be, taken along with development. In this view, the relative peace characterizing resettlement in Colony Para not only can be explained by the ambiguity of the socio-ecological terrain (relative stability in Gangasagar vs. spatial vulnerability along the long fragile shores of Bankimnagar, and availability of public water utilities in Gangasagar vs. effectively village-controlled water access in Bankimnagar), it also refers to future contingencies. Considering divergent experiences of resettlement between these two colonies reveals how the distribution of papers and the remaking of homes are placed in relation to other development trajectories. Similar to the unpredictability of the terrain, these futures are difficult to ascertain. Emphasizing political and ecological horizons comes as unsurprising in ethnographic accounts of (forced) mobility and settlement. Yet it carries essential analytical weight, I suggest, in addressing the thorny issues of thinking, and managing, ongoing and future flows of environmentally displaced people.

Terminating Unacknowledged Forms of Rehabilitation

Even if resettlement in Colony Para was a rather peaceful affair, in the eyes of newcomers, it still provided moments of quite intense competition around lots and loyalty—key currencies in quests for power in rural Bengal. Colonizing the swamp on the southern shore was a process of transplanting loyalties and networks from the eroding islet in the northwest down here. From very early on, competing claims were vocalized. Consider Svagat. He must have been an odd fit in the early days of Colony Para. A jack of many trades, in his youth he had joined a theater troupe that traveled the length and breadth of West Bengal. Returning to the island, he was still without land and means. When Svagat heard news of the colony, he sprang into action. He convinced the Gangasagar Gram Panchayat to hand out land to him, making him the first recipient of land in this colony who had neither lived on one of the islets to the north nor experienced environmental displacement. Years later, over tea in his roadside restaurant, he told me that he was, in fact, the architect of the colony. And that is true, in a sense. He may not have been the mastermind behind the settlement, devising schemes or arranging embankments or roads, but he left a mark by influencing its social composition. Mobilizing his rhetorical skills and the close contacts he enjoyed with political leaders around Gangasagar, he successfully pushed for the inclusion of destitute people from southern Sagar into the ranks of beneficiaries. Maneuvering the game of distribution and loyalty successfully—by proving allegiance to his seniors and winning clout among the landless by arranging for land titles—boosted his acumen and allowed him to ascend the ladder of local politics. In doing so, he also shaped the outlook of the settlement, undermining the special status of Colony Para as the priorities he set contradicted the widely distributed notion that Colony Para was earmarked solely for people displaced by an encroaching sea. In a way, then, with Svagat's interventions, what I call unacknowledged forms of relocation fed back into the trajectories of ordinary land distribution.

The shift is inscribed into Colony Para itself. Its last settled parts in the north and west were exclusively settled by people without exposure to coastal erosion. Over the years, good relations emerged between older and newer parts of Colony Para. Inheritance laws have diluted the differences in plots, giving way to a general sense of overcrowdedness and of having to make do with insufficient plots on meager soils.

There is one more way in which settlement in Colony Para intertwined the political fortunes of aspiring leaders. Along with a group of twelve families, all hailing from Botkhali, Nur, who showed me the lost cyclone shelter (chapter 3), built his home on the most haphazard site of Colony Para, the outer ring embankment facing the canal. He did so autonomously from the routines of land distribution and, initially, without state sanctioning. His move to settle here was arranged by a late political leader from Botkhali, in return for Nur's allegiance to him. Things became complicated. The leader died, with the influence of his party, the CPI(M), tumbling down with him. In this shifting scenario, their claim to a space on top of the embankment was under threat, by the shifting tides of political power as well as those of the erratic waters.

Ensuing governments proved to be little different in their attempts not only to rework the setup of ministries, but first and foremost to entrench themselves in local politics through the means of welfare distribution. The paperless, until then accepted settlers on the embankment, maneuvered the shift cleverly and eventually secured inclusion into the folds of beneficiaries of IAY. This meant the replacement of their modest mud houses with concrete homes paid for by the government. The buildings themselves terminated any immediate threat of eviction, rendering inhabitants as legible subjects of the state.

In one view, the shift from mud to concrete, from thatched hut to "proper" house, showcased the accretive ways through which settlements of the poor shift with time, making use of whatever means are at hand (V. Das 2011). In another view, the shift signals Nur's establishment as an officially sanctioned resident of Colony Para, and therefore his success in achieving legibility to the state, and paper-backed claims to permanence. In this case, there was no need for fiddling with papers, stored in the ubiquitous plastic folders away from thieves and vicious waters, to buttress the claim to state legibility. The letters affixed to the signboard on the door attested in writing what the structure as a whole already conveyed to most islanders: a signature investment of the IAY. More than offering a home set to endure weather, the house indicated legality in the inhabitants' claims to the land. If land ownership is performed through series of claims that involve material transformation, appeals to state officials and testimonies of pasts lived here (J. M. Campbell 2015), the state-sponsored building materially cemented such a claim. In so doing, it also rendered a form of habitation legitimate—a house on the public utility of an outer ring embankment—that was anything but, and lent a degree of security to a dwelling of utmost vulnerability.

The Perils of State Gifts and Shallow Roots

Across these divergent trajectories of settlement, one commonality remained: the principal unreliability of the land and titles. Doubts about the future of the land, or one's place on it, seem unevenly distributed. People like Juddhistir Jana, for instance, appeared free from them, positioning themselves as having finally landed a permanent space, removed from the vicissitudes of human threats through reliable papers, and of environmental threats by virtue of being secluded in the oldest parts of Colony Para. Many others, however, felt uneasy about their future prospects. There was, of course, the realization that mounting household sizes had to be fed by the land during ever-shorter periods of the year. Many worried that, if no measures were taken, more and more fields would eventually turn into homesteads to accommodate growing families. Such sentiments are, of course, shared across South Asia's land-scarce rural terrains. In addition, residents were apprehensive about the future of the settlement. I have noted in chapter 3 how the noises and rising water tables during the most incessant spring tides brought back memories of past displacement, fostering a sense of profound insecurity as the waters threatened to overtop dikes. Land seemed ultimately unreliable and its future up for debate. In the oldest parts, secluded from the waters by homes, and already-existing infrastructure, the situation was less precarious. For people next to or living on the dike, the situation remained tense, even if concrete houses seemed to lend a degree of additional security.

However, paper-based claims to land seemed porous themselves. Land distribution remains a vague and haphazard affair. Any land distributed through high-handed processes over which islanders felt they had little control could, as so many of my interlocutors feared, be taken from them at any point. This applied first and foremost to residents peopling the edge of Colony Para, tolerated as squatters like Nur, until they managed to secure state-funded concrete houses. Residents holding papers voiced similar concerns nonetheless. After all, I learned, the papers they held were of little use should the state have other plans for the land. Papers could be quickly annulled, and lands emptied, as people in Bengal know all too well given the recent past of land acquisition and forced eviction for development projects (Banerjee et al. 2007; Nielsen 2018). Not bought but bestowed, land and titles remained, to many of my interlocutors, a favor from the state. Consequentially, the land never *really* belonged to them and might be taken away at any point. The state gives and takes as it pleases, I heard people saying. Much like the river, I thought.

During my fieldwork, nothing concrete was on the horizon in terms of large investments demanding forced land acquisition. Yet, the development frenzy in this emerging tourist destination, and the obvious prioritization of pilgrimage traffic and bourgeois holidaymaking, might wash up a big project or investor anytime.

At the hands of the estuary and the state, life in the colonies was bound up in limbo, oscillating between trust in permanence and fear of ruin. Between these positions, specific relations to land emerged. These relations are deep and embody the poetics of dwelling, yet shallow in time and, often enough, uncertain. In this view, the forms of dwelling I have laid out here, imbued by the churn of displacement and emplacement, contrast with how settlement tends to be theorized in agrarian lifeworlds in the Bengal Delta and beyond. They neither seem to embody the rather stereotypical take on villagers as sons of the soil, enjoying deep roots and treading their environs as places brimming with pasts and potential. Nor do they resonate with takes on lives on the river islands characterized as enfolded in mobility as they dance with the river in hybrid arrangements between land and water (Lahiri-Dutt and Samanta 2013). Islanders deployed what I like to think of as shallow roots. Stories and relations to people and things had engendered a sense of emplacement, lives rooted in the land and the peasant futures it seemed to bestow. Yet in so doing, land on the islands tended to remain within the framework of *jayga*—a container, a base, or a space—and belonging was often voiced in reference to ancestral homes across the estuary on the mainland. Considering the relatively short period of habitation on Ghoramara or Lohachara, it remains to be seen whether Colony Para or, by extension, Gangasagar might eventually emerge as *bari*, an ancestral home. Any fears of being moved on by a fickle state are tempered by a firm belief in the safety of land hidden behind a dune and adjacent to a major pilgrimage site.

Beyond One-Size-Fits-All Solutions

Analyses of environmental displacement need to account for what I have called "unacknowledged relocation." This incorporates securing place, paper, or homes after environmental displacement through unrelated and rather unspecific forms of social welfare distribution. Addressing these barely visible forms allows us to correct accounts in terms of statistic distribution of resettlement as well as to assess testimonies and studies with a view on formulating necessary policy recommendations and guidelines. Moving in that direction, my account highlights two rather different trajectories of resettlement, one marked by conflicts, the other by what I call "sharing in the new."

If coastal worlds must face up to hard choices as we settle into the Anthropocene, we would be wise to take seriously how people navigate such changes both physically and bureaucratically. This is not the place to reignite the debate on how managed retreat replicates injustice deeply built into the procedures of environmental management, where some populations such as those in urban Indonesia benefit from dedicated programs, while many others are out of reach of assistance (Ley 2021). My ethnography ultimately straddles these perspectives, and the stories I depict here are—in all their misery—largely success stories. I foreground these not to introduce resettlement on Sagar as a shining example contrasting with India's overall abysmal record of resettlement. Instead they help me to carve out the experience of resettlement and its contingencies.

I have dwelled at some length elsewhere (Harms 2017) on the issue of citizenship on crumbling shores. Here I suggest that considering settlement drives that may have no bureaucratic link (in name or procedures) as relocation-in-practice may enlarge the scope of both actual resettlement drives and the material to work with in order to account for the experience of environmental displacement. Using the way rather normalized settlement regimes may be adapted so as to—at least partly—respond to the demand of environmentally displaced persons opens up a plethora of questions. One set of questions, of course, concerns who makes it onto the list of beneficiaries in times of scarcity. Another asks what the schemes do and how they come to be lived with. This chapter mainly engages with the second one. I use the tweaked or piggybacked land distribution schemes as a vantage point from which to reflect on the variety of replacement. This considers not only the politics of exclusion but is attentive to the affordances of terrain, the power of lists in generating subjects, and the advantages of placing oneself favorably in a climate of development. While these are as contingent and specific as ethnographic detail tends to be, they provide critical input in deepening the understanding of resettlement, both in terms of providing apt theory and by challenging the tendency for all-too-easy answers and one-size-fits-all solutions.

Chapter 6

Keeping Dry, Staying Afloat

The Politics of Coastal Protection

On the salty seams of the delta, abandonment often takes the appearance of a construction site. Jana's house is a good example. During my last visit, when we step out of his garden, it is only a few meters before reach the fortified shore, where a simple mud embankment safeguards the houses on Ghoramara's northeastern flank from the estuary and the brackish waters it carries. This outer embankment is as new as it gets, and is already disappearing.

We step on a mound of freshly piled-up mud. It has only been fifteen days, Jana tells me, since this stretch was completed. It is raw mud, without the enforcement of a wooden frame or sandbags. Rains or the baking sun have yet to smoothen its top or interior. The outside, however, is already under heavy attack from the tides and waves (see figure 6.1).

"It won't stay, it will go this year," he says, a sense of despondency in his voice, the efforts of raising the mound not forgotten. This, and the most likely fleeting presence of this embankment section, is echoed in the subtler messages the material environment sends. Further out, another embankment barely reaches above the water; it was erected only last year, but less than half still exists.

Not many people pass through this far corner of remote, tiny Ghoramara, but many leave. A few weeks earlier, Jana's neighbors moved away for good. Their home sits empty, with black holes where windows and doors once stood, and a garden left to the looming waters. More of such ghost houses dot the vicinity. Some of them are freshly forsaken, standing in wait as if their owners were to return any moment. Of others, however, only remnants of outer walls stand, poised to collapse into the gurgling waters licking at their feet.

I join Jana as he inspects the frail attempts meant to help him weather the waters' approaching his home; the challenges of living with often-weak or defunct embankments become quickly palpable. How islanders

Fig 6.1. A new and already-disappearing embankment, Ghoramara Island. Photo by author.

populating the immediate coastline navigate the tensions of utter dependence on coastal fortification, the certainty of disrepair, and the ultimately futile maintenance of built protection measures permeates this chapter.

Embankments are vivid social sites. They do not simply stand between water and land; nor are they simply material forms of efforts to preserve life on land in these otherwise amphibious tracts. In this chapter, I turn to coastal protection assemblages and explore their diverse constituents. Approached ethnographically, this opens a vista into the politics of environmental governance and socio-ecological relations. I call attention to their structuring and restructuring of life on the islands, to their many constituents—including mud, concrete, plants, and gestures—and explore emerging tensions.

Embankments, of course, are not uniform. Simple, enforced earthen dikes coexist with sandbags, the occasional brick enforcement, or huge mats interweaving boulders. In very rare cases, concrete seawalls guard critical assets. It takes only a short walk to go from a zone of abandonment marked by substandard, improvised embankments, as can be seen from Jana's house, to one of rather sophisticated, capital-intensive protection interventions. This multiplicity only partly draws on site-specific differences. Exploring the uneven and layered nature of coastal protection infrastructures, and their manifold and partly conflicting constituents, is vital to any account of erosion and submergence on and around Sagar. Landscape transformation, ironically, also lays bare how such top-down approaches fare as they hit the ground. This chapter uncovers how coastal protection actually involves much more than mounds of earth or neatly lined slabs of concrete, and how they emerge out of the sometimes concerted and sometimes conflicting effects of what people do and of how materialities fare.

Most infrastructure enables the circulation of stuff, people, or ideas by being firmly built into place (Larkin 2013). To the degree that infrastructures are concerned with flows, they also are means of arbitrary routing. Not everyone is connected, nor is it all evenly distributed or to be found everywhere (Graham and Marvin 2002; Gardner 2012; Harms 2019). Coastal protection assemblages emerge as an extreme case. Alongside, for instance, militarized border infrastructures (Carse et al. 2020), they form a class of interventions into the socio-material fabric that obstruct flows and guide them, whenever desirable, through closely monitored gateways, such as sluice gates or pumps. Flow, then, appears as the adversary in much of modern coastal protection design, and neat distinctions an ideal aspired to.

In thinking through infrastructures along Sagar's shores, I draw on recent accounts that highlight their liveliness. This makes me wary of accounts positing them as purely technological artefacts or mute things. As with all technical devices, infrastructures require constant labor to make them work in specific contexts. This is a labor of translating and anchoring that shapes devices as much as the human actors involved (de Laet and Mol 2000; von Schnitzler 2013).

To speak of lively infrastructures is to speak of more-than-human entanglements. Ashley Carse (2014) demonstrates that the Panama Canal requires massive freshwater input from adjoining rivers and streams. Without it, ships could not move through the canal. Water supply turns out to be an essential facet within this infrastructure, and its day-to-day operations involve the implementation of dedicated governance procedures geared toward securing water provision. In this view, water, forests, and their managers are drawn into the fold of infrastructure, and the borders between these actors are as ambiguous as they are porous.

Turning to another shoreline critical for the functioning of global capitalism, Stephanie Wakefield and Bruce Braun (2018) examine recent attempts to protect Manhattan from surrendering to sea level rise by cultivating oysters as natural protection. Citizens toil not only with reintroducing an almost extinct species to its original habitat but with cultivating "living infrastructures." Life in this case is harnessed doubly. These infrastructures consist entirely of organisms (and their residue), and are hoped to grow with and along the rising seas.

In this chapter, I build on these works and consider coastal protection as lively infrastructure. Instead of thinking of it as a thing, or an outcome, I explore coastal protection as an assemblage that comes into being through heterogeneous constituents, and as something that evolves through myriad interlocking processes and practices, rather than standing in place. Not an end in themselves, I consider how these assemblages operate as instruments of landscape transformation.

To do so, I first turn to the political dimensions of embankments as instruments of landscape transformation. The variety of embankments and their embedding found on Sagar and Ghoramara—some low-tech and fleeting, others grandiose and more durable—cannot be sufficiently explained by divergent geomorphological conditions demanding more protection in one place and less in another. This foregrounds, I argue, an unevenness of provisions, while shedding light on the politics of state care and "uncare" (A. Gupta 2012), and the lopsided mosaic of protection measures that islanders need to navigate.

Speaking of instruments of landscape transformation might be seen as an exercise in techno-optimism and grandeur. Projects of embanking swamps or carving land out of marine terrain continue to be animated by such, like in the making of Holland or of contemporary coastal reclamation projects in Indonesia (Keller 2023). Problematizing the imagination of progress and duress, and the futures they are pitched to enable, I call attention to the ephemeral nature of these interventions. After all, embankments may quickly be overrun or undercut, producing ripple effects in coastal dynamics not fully understood. They are united in the likelihood of their demise, which renders them instruments that slow down rather than stall, that temper loss rather than gain new land. In this view, temporality and the certainty of disrepair and demise, a dynamic I explored in conceptualizing erosion as distributed disaster (see chapter 3), comes into the fold once more. This time around, however, it is a feature of governance on these wettened delta seams, and therefore a politics more immediate than the matter of Anthropocene shifts in sea levels, riverine currents, waves, or winds.

Exploring embankments, however, requires looking beyond the schemes and plans devised by engineers and politicians. It requires taking note of maintenance. At the wetted seams of this delta, maintenance puts into view how coastal protection infrastructures serve as a bone of contention in cash-starved and underserviced lands, and how, by extension, the very frailty and immediate disappearance of the vast majority of embankments serves as a source of income. I explore how vulnerability and survival are intertwined, and how islanders navigate the openings and potentialities provided by imminent breakdown.

Finally, I depart from a narrow view on coastal protection infrastructures and consider the infrastructural provisioning of non-embankments. This entails observing mundane efforts of caring for threatened homes or plastering flailing embankments during moments of heightened risk. Doing so complicates the account of materials, practices, and persons understood to keep embankments in place. It foregrounds the role of embankments as they disappear as means of generating knowledge on a landscape in flux. Thinking unacknowledged and often invisible labor, together with the role of earlier embankments for coming to terms with an amorphous waterscape largely devoid of markers, calls attention, I will argue, to overlooked constituents and effects of coastal protection assemblages. Attending to such "minor infrastructures" helps to give agency back to resourceful islanders making interventions into the fabric of the coast.

A Patchy Mosaic and Instruments of Landscape Transformation

"Dams straddle and embankments," notes Amita Baviskar (2003, xi), "straitjacket most Indian rivers." The tattered ends of the Ganga are no exception. These instruments are pierced, their effects patchy and uneven. Along Sagar's eroding edges, a language of firmness and control fails, even if—and this is how I intend to use the metaphor—the sense prevails that drastic interventions are needed in order to tame a patient that has become a threat.

Giant slabs of concrete, patched around the southwestern corner of the island, mark such a drastic intervention. A giant seawall, gleaming in an industrial lifeless gray and towering high above the hamlet sitting in its shadow, is a recent attempt (see figure 6.2). Workers carved the year of its completion into the concrete as it was drying: 2018. On the wall, one year later, I speak to Ganesh Maiti, who echoes what others have already told me. "It will last," he says, "for now. For ten years we are OK." Temporality is writ large, even in this feat of engineering. Maiti's assessment highlights the imperfection of this boundary, its impermanence even by human standards.

Fig 6.2. Seawall, Lighthouse, Sagar Island. Photo by author.

Villagers like Maiti cherish the respite the wall offers. It replaces a notoriously instable and troublesome outer embankment, patched together out of wood, mud, and sandbags. Waters and winds gushing around this corner of the island would frequently pierce it, leaving homesteads drenched and shores eternally shapeshifting. On one end of the seawall, remnants of the older embankment still stood when I visited last, as if to serve as a second row of protection. On the other end, however, the older embankment reappears. Mounds of sandbags heaved upon mud and wood, again, act as the outer ring embankment.

With no visible gap whatsoever, the shift from concrete slabs to sacks of sand clearly marks a rupture. Cheaper materials, built to a height significantly below that of the seawall, zonally mark the end of investment in coastal survival. Coastal protection regularly relies on diverse types of materials, designs, and approaches, assembled within the constraints of ecology, finance, or expertise. The particularities of such assemblages lay bare, I argue, the grammar of coastal governance and the state care/uncare they perform.

While such high-tech arrangements appear standard in many extremely low-lying, densely populated delta tracts (Kane 2012; Ley 2021), they continue to be both an exception and a spectacle in the Sundarbans. Many villagers I spoke to, including Ganesh Maiti, held the port agency responsible for constructing the seawall. The port authorities had indeed imprinted themselves deeply into the area, which is dominated by the colonial-era lighthouse. This building was, until the arrival of mobile towers, by far the highest construction on the island and still dominates the flatlands making up its southwestern edges. As a consequence, islanders refer to the village spreading around the bright red tower simply as "Lighthouse." They knew full well that port authorities had invested to secure the asset this remote outpost was.

A few years ago, updated marine navigation technologies rendered the lighthouse redundant. Given its age—and its key role within traffic between Calcutta and the wider world within this period—it was designated as a heritage site. It isn't the first lighthouse, however. There was a predecessor, which the port had to abandon to the waves in 1905. The paper trail it left resonates with the temporal envisioning both Maiti and Jana put forth.

In the first decade of the twentieth century, lighthouse staff regularly notified port authorities about coastal erosions' advances. Their reports included measurements of how many feet of the embankment were razed in a given year (see table 6.1). Alongside the paper trail around decisions to abandon the old and build a new lighthouse, these documents are among the oldest ones I could locate showing the effects of coastal erosion on settled parts of the island. While they prove that coastal erosion has been rampant long before climate change made its presence felt, they also show temporalizing takes on their health and reliability by people on the ground. Above all, they demonstrate the long history of variegated coastal protection, connected, among other things, to investments by the port.

To this day, the lighthouse area continues to be a vital element of the port's communication infrastructures and cannot be abandoned to the

Table 6.1. Documentation of coastal erosion based on embankment measurements

Date	Total Erosion	Erosion in the Reporting Period
30 September 1909	525 feet	washes away 11 feet
1 December 1908	566 feet	washes away 74 feet
1 December 1907	640 feet	washes away 60 feet
1906	700 feet	washes away 32 feet
1905	732 feet	washes away 59 feet
1904	791 feet	washes away 55 feet
1903	846 feet	washes away 124 feet
1902	910 feet	washes away 34 feet
1901	1004 feet	washes away 31 feet
1900	1035 feet	

Source: Excerpt from file "Erosion on the Bank of Saugor Island," Marine Department, Marine Archives, Kolkata.

waves. It is a barging point for boat pilots who guide commercial vessels through the treacherous river between docks and sea. A jetty, a rest house for river pilots, and working road communication to this outpost at the very confluence of river and sea remain critical for the overall functioning of cargo handling in East India's main port. Constructing the concrete seawall was simply the latest in a long line of attempts to temper the coastal erosion threatening port property, and merely another instance of interventions into the texture of the waterscape undertaken to protect port assets and traffic between the city and the bay.

Not all islanders agreed on the port authority leading the construction. Conversing on a bench situated right next to the place where sandbags reappeared, villagers called the concrete colossus *Aila bandh*. The name highlights a politics entangling disaster response and coastal protection infrastructures with the fate of Bengal's ruling party. *Aila bandh* signify embankments, *bandh,* built in response to Cyclone Aila, which ravaged parts of the state in 2009. *Aila bandh* are not limited to this bend. They have come up in several places across the delta front (Basu 2020)—so far, however, not on any of the islands I did fieldwork on.

As comparatively large-scale interventions scattered across the delta, *Aila bandh* are spectacles of care, and of technological and financial prowess, in a landscape otherwise known for being impoverished and lacking services. How those of my interlocutors who understood the sea wall to be an *Aila bandh* explained its existence here, on this particular shore, further accentuated the spectacular. "Didi gave it," they said. "It's her gift." *Didi,* literally "the elder sister," is a popular moniker for West Bengal's chief minister, Mamata Banerjee, in office since her 2011 landslide victory.

Since its inception, Mamata Banerjee's government skillfully used infrastructure development to win allegiance across the state. Using such gifts as bargaining chips to reward followers and to gain foothold in virgin territory is not specific to her rule but a cornerstone of electoral politics across West Bengal and beyond. Infrastructural provision here showcases efficiency and addresses specific demands through the interventions themselves or the distribution of funds so as to win over, and sustain, targeted voter populations. The divisive dimensions of such politics—enfolding some at the expense of others, and granting rights in a language of gifts—have been remarked upon often (see, e.g., Chatterjee 2011). Building on such takes, I highlight how investments into seawalls, and their portrayal as gifts, underline the uneven politics of coastal protection. This particular bend, and its village, received the gift; others did not.

Regardless of whether the port or the state's ruling party brought the seawall here, the outcome was in line with the comprehensive approach—the master plan, at least for the acutely threatened sections of the islands—that politically vocal islanders have been calling for. Streamlining substantial funds, bureaucratic efforts, expertise, appropriate materials, and technology, it seemed the best possible solution to counter the threats engulfing the island. Having a wall here, and only here, alludes to a politics of uneven provisioning. In a way, port authorities and state government appeared driven by a similar logic: protecting specific shores through resource-heavy interventions, thereby inviting conclusions on what was worthy of protection and what was not. Coastal protection, in this view, is not only provided unevenly, it brings about a mosaic of differently enabled efforts carrying far-reaching consequences for people and ecologies.

Debating the reappearance of sandbags, and therefore the return to a much less sophisticated and—according to islanders—much less reliable form of embankments, brought this to the fore. Susceptibility to surges and erosions had increased, there was little doubt. Obviously, saltwater intrusions, and their destructive effects, are much more likely on terrain adjacent to the fragile mud embankments enforced by sandbags and bamboo than on terrain nestled behind seawalls (see figure 6.3). Whether the seawall will live up to these expectations, and whether it will serve as a protective measure, at least for the next few years, as Ganesh indicated, remains to be seen. In any case, its predestined future

Fig 6.3. Embankment made of sandbags, wood, and mud, Sagar Island. Photo by author.

demise spells out the risks flowing from the decision to build the construction to this point here and no farther.

My interlocutors explained that the seawall was originally meant to be longer and encircle the exposed nose of land situated between the vast Hugli estuary, the open sea, and the outlet of a channel draining the island. But construction halted prematurely. Villagers living here, they told me, had been unwilling to give up their land and make way for the wall. On one level, abandonment and mistrust wove into one another. On another, this exemplifies the exact kind of maneuvering the politics of gift giving embodies, one of specific negotiations between local leaders and situated publics.

I do not want to go into detail on the accusations of resistance and the politics they engender. Instead, I want to use this figuration—the seawall's sudden end and the allegiances surrounding it—as a lens into the continuous reshuffling of coastal vulnerability.

The seawall comes to a fairly abrupt end. While it does not include further elements of advanced technologies of, say, coastal land reclamation, it could be argued that the area behind the seawall has been deemed worthy of full protection, or at least the greatest level of protection possible. Yet somehow, the adjacent terrain comes to be encircled by an embankment that is much less sophisticated, though still—by local standards—rather durable. I like to think of this area as leaning toward abandonment. In other words, both appear as points on a spectrum, their positions bound to budge with the ecological and political shifts of the coming seasons, local politics, electoral cycles, or the fortunes and ingenuity of engineers. The shift in coastal protection types indicates a hierarchical ordering of territory. Specific shores are safeguarded, while others are deprioritized, the subject only of low-tech interventions. In both cases, and this is the point I want to stress, shorelines are made and remade by technocrats according to their priorities and constraints, inscribing and reinscribing these into a landscape sustained by patchy interventions. Affected islanders negotiate landscapes shaped by these decisions, and are left to cope with both anthropogenic shifts in the waterscape and the much more direct consequences of protection measures.

Slowing Down: On Failure, Withdrawal, and Abandonment

As with all major interventions into the complex hydrology of the delta, it is much too early, perhaps ultimately impossible, to ascertain how the seawall will fare and impact the waterscape. The history of the

Sundarbans at large, and of the handling of marine traffic in particular, is ripe with failure and frustrated hopes. Consider the devastation of the new port in Canning (Sarkar 2010), abandoned after a cyclone had struck and unexpected siltation halted all traffic. Or recall the collapsed cyclone shelter explored in chapter 3. It seems justified to understand the seawall as yet another experiment rather than a definitive intervention. At the very least, such a take explains the hesitant hopes attesting nothing more than a few years of safety.

Closer to home, and much less dramatic than the spectacular failure of Port Canning, Ghoramara has seen its own failed attempts at halting erosion. Once again, the port had tried to safeguard its assets and enable marine traffic. This time, the focus was on reducing the siltation of the port channel, which has plagued the port for centuries. The high amounts of sediments present in the channel tend to sink and amass, raising the bed and thereby reducing the depth. This threatens to block the harbor for the large cargo vessels that have become global standard (Carse and Lewis 2016). At the same time, siltation is uneven. Sediments tend to amass into irregular chunks and bars, rather than even layers, regularly causing short-lived obstacles in the channels.

Siltation is common, not surprising in a hydrological setting as complex as the active delta. Geographers identify settlement of the delta as a major factor here (Dewan 2021b). If presettlement the tides would wash over islands, leaving sediments as they flushed out, embankments block them from doing so, increasing the amount of sediments in estuarine waters. Simultaneously, riverbank and coastal erosion feeds into these processes, adding even more sediments to the waters close to the mouth of the river. Tidal dynamics, finally, hinder the flushing of sediments out into the ocean. Taken together, these processes render the Hugli estuary a "sediment sink," with sediment rates rising across the last couple of years (Bandyopadhyay et al. 2014), only intensifying what to the port appears to be a sediment crisis.

Port authorities seek to counter this chronic issue in a number of ways. Dredging continues unabated, as does the perusal of river pilots circling between Sagar's Lighthouse and the docks. In addition, engineers devised plans for laying subsurface walls at Haldia, as noted above, which began to direct tidal currents toward Ghoramara. The result was a dramatic increase in erosion on the island, which eventually prompted port authorities into action. In 2009, the Kolkata Port Trust (KPT) cleared funds for an innovative measure, both large-scale and flexible enough to bring relief. Contracted companies assembled concrete into boulders and wove them with durable ropes into giant mats before anchoring them on the shore (Ghosh, Bhandari, and Hazra 2003). Once ready, those mats were supposed to cover the area from the high tide line (and thus they needed to be fixed to embankments in most places) all the way below the low tide line. Bouldering—as islanders began calling it—was meant to cover the majority of Ghoramara's western flanks. These were massive works.

Islanders welcomed the throng of activity. After years of empty promises, authorities seemed to swing into action. At last, the crumbling shores were enveloped in a humdrum of construction much more ambitious than the previous repairing of razed mud embankments. Many islanders found unskilled labor jobs. More importantly, perhaps, the massive investment signaled that state agencies finally took interest, beyond lip service, and concentrated funds. The sheer size of the operations signaled that not all was lost, and the durable materials, crystallizing in spectacular giant mats, underwrote hope into the island's persistence, its future.

Contrary to these visions, the subcontractors involved in implementing bouldering showed little interest in the continued existence of the island or in upending coastal erosion. While on a walk with me, one subcontractor made it unmistakably clear that bouldering was not done to safeguard the island but to improve the channels of the port. Or, at least, to slow down their further deterioration. The port leadership, he explained with a whiff of authority, saw the drastic erosion on Ghoramara as impairing the condition of the waterways: land would continue to be submerged here in the mouth of the Hugli and encase it even further. In this instance, the interests of port and islanders aligned to a degree, for the strengthening of the shore reduced land loss. But even so, the boulders represented a self-interested move by a corporation known for its long history of intervening in the waterscape only to meet the demands of globalized trade. The well-being of the river or its people was immaterial.

A few years later, these hopes were gone. The mats had been unable to stop the waters' destructive effects. Underneath, erosion had continued. Slowly, boulders were coming loose. With their impending demise, erosion would kick in, islanders feared, with full force again. Worse, after bouldering was completed and contractors packed up, all interest seemed to evaporate. The usual drill of watching embankments being built only to have them razed in no time continued unabated on many sections of the shore, just like it did in Jana's hamlet.

The parallels between the bouldering initiative and the seawall are striking. Hopes, however, seem more hesitant lately, involving merely a sense of being safe for now rather than forever, at least to people like Ganesh Maiti. Similarly, laying boulders and constructing sea-walls—contained to sites considered relevant by external experts, and bringing them under specific protection—spells out mosaics of uneven coastal protection.

The techno-optimist grandeur invested in both interventions now seems severely fissured. Both forms of high-tech, weaponized coastal protection infrastructure still seem ephemeral. Now that hopes in bouldering have faltered, sounding a warning call for what comes next, even such weaponized interventions appear to offer mere moments of respite. For ten years we are safe, for ten years we are OK. Even the most advanced forms of coastal protection appear to provide little more than a reduction in the speed with which the salty waters enter islanders' lives. They enable a cushioning of processes that appear—within the order of things—impossible to upend. If we think of embankments as instruments of landscape transformation, the instances considered so far seem to be severely limited: reducing the power of onslaughts without ever halting them, slowing the speed of erosion without countering it or turning it on its head, and enabling provisional futures instead of assuring permanence.

Contrary to the implicit assumption that embankments were stabilizing and advancing claims into the ocean, on Sagar they were often enough means of slowing down an advancing estuary. Similar can be found across the whole of the Bengal delta (Aditya Ghosh 2017; Dewan 2021b; Jalais and Mukhopadhyay 2020). The work of embankments, as instruments of landscape transformation, is therefore frequently a negative one. In the way embankments are designed, funded, and implemented, protection and abandonment are ultimately difficult to disentangle from one another. I explore these entanglements in the next section.

Abandonment

As if foretelling the seawall, Partha Neogi never failed to call for concrete embankments to safeguard his native village. His home, however, used to be on the other side of the island, on its rapidly eroding southeastern corner, in Botkhali, a village whose future appeared—if looked at through the prism of the collapsed cyclone shelter (see chapter 3)—increasingly sealed. Partha, however, did not give up. In 2009, when we first met, he was fuming. A quick-tempered street-level bureaucrat, he filled a post at the lowest rank of the state administration, the Gram Panchayat, and sought to use whatever influence he had toward securing meaningful protection measures. Raging at the lost opportunities of protecting homes and fields, and the futures awaiting him and many of his fellow villagers, he called for a comprehensive solution. Nothing else would do, he insisted to the small group of nodding shop owners, pharmacists, and farmers that had assembled in front of his modest hut. What was needed was, and here he used the English term, a "master plan." It needed to be different from the typical repair works and to mobilize scientific expertise. He had talked to a team of foreign engineers, who came for a visit to assess this hotspot of coastal erosion. They had assured him, he went on, that saving the village was technologically feasible. But it needed a master plan—he was hammering home his point—and this was to be based on the lavish application of concrete.

Over the years, he grew more desperate, as his gripes led nowhere. The government invested in a seawall at Lighthouse and erected a state-of-the-art cyclone shelter in the interior, a short bike ride from where we first met. It wasn't as if the government's hands were tied and the means to grant protection were unavailable. They just didn't seem to care.

In his village, however, the government resorted to funding only the least expensive and least durable types of outer ring embankments people knew: sinking wooden poles at regular distances from one another into the ground, as an enforcement, and heaping mud onto them.

While such simple embankments seemed more or less sufficient in less-affected areas, such as Colony Para, people in Botkhali had long lost all confidence in them. Villagers had seen them built only to get razed with the coming monsoons and the heightened waters, soaking rains, and rough winds they brought. One after the other, year after year.

I will detail below how embankment maintenance provided livelihood to villagers able to secure shifts in the officially funded work gangs, and how the recurrent nature of the task of re-erecting razed embankment contributed toward sustaining families. At this point, it is worth emphasizing that villagers felt abandoned as they saw the same measures being replicated time and time again despite being proven unfit for the task. Bitter complaints about the inadequacy of state interventions echoed throughout the ruined villages along the sinking shores. People knew newly completed works to be already inadequate. Along specific stretches of the shore, from Jana's neighborhood to Partha's devastated village, abandonment took the form of provisioning.

In his study of humanitarian assistance, Robert Redfield (2012, 180) highlights the last-ditch efforts of interventions. These interventions target the mere survival of bodies at risk, rather than the realization of human rights to, say, health and dignity. Such "minimalist forms of care" tend to appear through institutions oscillating between philanthropy and corporate interests. They seek to alleviate situations where overburdened or dysfunctional states fail to live up to their most fundamental duties, such as providing clean drinking water to children in sub-Saharan Africa. I find this notion helpful in conceptualizing the specific form of structural violence at play in this scene of abandonment. Rebuilding unsuitable embankments, and thereby continuing to invest in inapt measures, is not exactly the same as the exclusion of eligible beneficiaries of state welfare or the mere existence of development projects on paper (N. Mathur 2015; A. Gupta 2012). Nor does it simply stick to the standard, since standards in coastal protection vary widely from region to region, and from stretch to stretch of the same shore. It rather involves a modality of provisioning that sticks to the bare minimum, aimed at securing modest survival while bypassing human and citizen rights, and it showcases an institution at work.

The minimalist and insufficient nature of low-cost coastal protection does not go unchallenged. Partha is, of course, not alone in lamenting government inaction or the faults of this or that political party. Others voice their frustration with ultimately disinterested, hapless bureaucrats failing to face ground realities as they assess the situation—or so a widespread trope goes—floating along on their inspection boat rides. Local administrators, on the other hand, would rebut such claims, and partly blame empty coffers for the sorry state of protection. Some even go as far as blaming mischievous villagers for looting building materials and weakening works (Aditya Ghosh 2017).

The minimalist and far-from-sufficient character of low-cost coastal protection comes strikingly to the fore in bureaucrats' decision to retreat the coastline—that is, to forsake a razed outer embankment for good and construct a new outer ring embankment farther inland. If coastal erosion involves, as I have suggested, the undoing of a given terrain's hospitability, the withdrawal of an outer embankment to the interior amounts to a turning point in this process. The land—including homes, water pumps, graves, and by then mostly already dead fruit trees—comes to be exempted from state protection and abandoned to the sea.

Decisions on when to shift the embankment, and how far, are taken by the local and little-loved Department of Irrigation and Waterways. Consultations between engineers, bureaucrats, and villagers on this matter could certainly provide for an interesting discussion of infrastructure in the making, were it not for the secretive handling of these matters. To villagers, the rhythm of shifting embankments remains elusive. For as long as I have known Botkhali, embankments have been shifted every four to six years. The exact number of years any embankment was upheld is an indicator of the velocity of estuarine waters' attacks and of state abandonment. Razed by waters and shifted by governments, embankments appear as instruments of landscape transformation operating less by securing terrain and more by merely slowing down its watery demise. Funding priorities and hydrological trajectories intersect.

Bureaucrats often choose one of the ubiquitous minor interior embankments as the new outer embankments, and have workers transform them. Sometimes new embankments are laid from scratch. Either way, those living behind what has been decreed the new outer embankment now have the tides reaching closer to their gardens and fields. This means an increasing likelihood of floods, as well as an overall increase in salinity in their gardens, hampering crops. For these reasons, the withdrawal of the dividing line between protected and sacrificed land, between landlike and liquifying soils, between still somewhat fresh and full-blown salty terrain unsurprisingly involves anxieties and tensions.

At the same time, the timing of withdrawing the outer line of defense is in itself an indicator, similar to the paper trail left on the destructions of the Lighthouse embankment noted above. The rate at which the administration abandons former ring embankments and resorts to new ones feeds into islanders' assessments of the severity of the situation, and of what the future holds for people living within terrain situated on what appears to be the pathway of salty water pushing into the land.

So far, I have looked at embankments through the lens of governance, and explored the contexts and ramifications of decisions on sites, types, and timings of specific coastal protection infrastructures. In the next section, I return to Colony Para. To complement my account of engineers' decisions and their repercussions in the terrain, I now look at tools and embodied practices and how matter such as salty mud or timber comes to be molded, and how all of them coalesce to form a politics of doing coastal protection on the ground.

Maintenance

Mud embankments need to be overhauled regularly. Even if in most places around the island this happens only once every couple of years, and not every season as in Botkhali, the expenditures and thus the income to be generated are substantial. Since this was a recurrent task, maintenance became an important source of income for chronically impoverished populations living on the edge of the islands. Here a paradox emerges: embankment maintenance as an embodied practice reveals how embankments are, precisely due to their very fragility and low-tech appearance, critical sites of livelihood generation.

On today's Sagar, in fact, embankment maintenance sucks up most of the funds earmarked for rural infrastructure—since roads generally also feature as embankments, there is a doubling of connective infrastructures (roads) and blocking ones (dikes), and they regularly head rural development expenditure lists. Once sanctioned by the Department of Irrigation and Waterways, concrete works are arranged through the Gram Panchayat, which employs workers, oversees tasks, and distributes wages after task completion. At present, embankment maintenance is routed through the procedures of the Mahatma Gandhi Rural Employment Guarantee Scheme (MNREGA), locally referred to simply as *eksau diner kaj,* "hundred days' work." One of India's flagship rural poverty alleviation programs, MNREGA pledges to provide at least one hundred days of paid work (as an unskilled laborer) to at least one member of any family seeking employment.

Based on empirical inquiries, scholars and activists argue that the claim to provide work to at least one member per family is hollow (Corbridge et al. 2002; Right to Food Campaign 2007; Adam 2015).[1] My empirical data confirms that far more people seek inclusion into the workforce, and that competition for shifts is a source of contention.

Furthermore, party-affiliated trade unions control the distribution of shifts, demonstrating how inclusion into the workforce, and thus access to welfare benefits, is a vehicle for enforcing party loyalties, and of gifts rather than of rights. Against this background, coastal protection assemblages are not simply uneven interventions into the socio-material landscape, but their continuous emergence—through repair, retrofit, or ruin (Howe et al. 2016)—simultaneously appears as the retrofitting of the polity.

This, of course, is not to indicate that this was a mere top-down remaking. Accounts of political transformation and unrest in West Bengal have shown how grassroots concerns may animate statewide shifts (Mallick 1993; Nielsen 2018). Adding to these, I show how marginalized actors use top-down decisions to transport themselves into the poli-

---

1. MNREGA also appears steeped in the troublesome legacy of food-for-work programs, which deal a blow to the dignity of people unfit for grueling manual work and in which earnings are insufficient; they also reinforce gendered and generational hierarchies in families (S. Sharma 2001).

tics of maintenance on their own terms, and in doing so further the paradoxical relation between disrepair as a threat and disrepair as income.

One day, I joined Aditya and his co-workers engaging in *mathi kathi*, literally "mud cutting," as this type of work is locally called, in order to overhaul Colony Para's outer embankment bordering the swamp. The task was simple but backbreaking.

Aditya starts by moving down the embankment on the sea-facing side, sliding more than walking down this slippery slope. Now that the monsoons have finally arrived, the temperature has fallen. Moving, however, has become more difficult in the freshly wettened mud. He does not stop at the base but continues on to his hole, where he puts down the empty container that he carries on his head. Now, as the day progresses well into the afternoon, the hole is quite deep already. Aditya steps in, takes up his hoe, and cuts pieces of the loamy, soaking-wet mud before heaving them onto the container. With two to three cuts, the container is full. He climbs out, lifts it onto his head, and climbs up the embankment. Finally, he tips the load on top of the old embankment, before sliding down again. Over the last few days, he and his colleagues have added another meter onto the embankment. The old structure is invisible now, buried under layers of chunks. All members of the gang work their own holes, and late in the afternoon, when the supervisor returns to check and register their workloads, the embankment itself, now enforced and heightened, appears much like an unkempt animal, waiting for the chunks to settle and be smoothed out by feet, water, and sun.

Cutting mud to maintain a state-owned outer embankment inevitably means digging holes in the salty, unsettled terrain on the river-facing side. The budget does not allow them to bring in earth from elsewhere, despite many islanders believing salty mud to be of an inferior quality for construction. It is frequently thought to be not as sturdy as mud from freshwater, which to them partly explains the frequency of embankment collapses. Many others hold the saltiness of the mud to be irrelevant and the mud cut in the swamp off the embankment as good as any. In either case, workers abstain from taking mud from inside the embankment as this is private land and put to intense cultivation wherever and whenever possible.

But to use the mud from the other side brings with it an overall weakening of the protection structure. The area is now checkered with holes, just like the one Aditya dug today. One per person per day. With the to-and-fro of the tides, any holes dug refill after some time with sediments. It is debatable whether cut holes change sedimentation dynamics to the detriment of the embankment, particularly as the sediments needed to fill it up cannot settle elsewhere and might produce mounds. But in many cases, taking mud from the amphibious terrain on the other side of the embankment means interfering with the plants and trees present there, which in itself may send ripple effects for the overall protection.

As noted, daily workloads consist of cutting rectangular chunks out the mud, measured in inches according to width and depth, and piled up to make the new or strengthened embankment (see figure 6.4). In order to dig holes, this gang, *any gang*, ends up cutting mangrove trees. They need to do so, they say, in order to clear the ground, to open up the space for cutting holes, or to clear paths leading toward the embankment. The strict rule issued by the Gram Panchayat, to not cut or harm mangrove trees because of the safety they provide against storms, surges, and erosion, is obviously suspended. One might argue, as Gram Panchayat officials do, that there is no other way to generate the mud. Others suggest

Fig 6.4. Holes dug for embankment maintenance, Colony Para, Sagar Island. Photo by author.

that the mangroves will grow back quickly, or that this is a form of corruption. Either way, for the families of laborers, it is a windfall. Islanders depend on wood for cooking. However, as almost all mangroves have been cleared, and all are currently protected, it is hard to come by. Islanders buy it on the market in stacks. As the work gang proceeds, maintenance workers have members of their households, mainly women, standing by to claim the wood. Accompanying the digging, hauling, and piling of the paid workers, they gather trunks, branches, and loose roots out of the dirt in order to cut them into pieces before finally rushing the fuel back to their houses. Thus, the people lucky enough to secure work profit doubly from the frail embankment.

Laying embankments as an embodied means of securing terrain, which included erasing trees and mining and piercing swamps with holes for the sake of lifeforms characterized by freshwater agriculture, ultimately clashed with official approaches to leave remaining swamps untouched, both for the sake of nature conservation and for coastal protection purposes.

Embankment maintenance also reveals the contested character of mangroves. Across the Sundarbans, some take tidal forests to be rare instances of untamed nature on an increasingly denuded, quickly urbanizing subcontinent. To others, they are sources of income, food, and belonging. All share in the conviction that mangrove forests efficiently shelter settlements from the ravages of the sea. Such notions have deep historical roots. Colonial officers, for instance, long emphasized the need to keep mangrove belts as protective layers between newly established agricultural lands and the bay (Dewan 2021b). Survivor stories, swapped in the wake of contemporary cyclones, affirm this take (Statesman News Service 2021). Cutting-edge research, and science communication seeking to bring this to wider attention, regularly underline the many benefits mangroves hold for coastal communities navigating increasingly rougher seas (Krishnamurthy 2023). In this view, mangroves are critical constituents of coastal protection infrastructures, playing a part in greater assemblages enabling land-based forms of life on the lowest-lying shores. In a recent nomenclature, mangroves feature as "soft protection" that ideally complements "hard protection" in comprehensive approaches to coastal management (Gesing 2016).

Cutting mangroves in order to build dikes involves what Lauer and his colleagues (Lauer et al. 2013, 41) call "resilience tradeoffs." Investing in one form of protection means neglecting another. The robustness of one measure (the embankment) underwriting resilience was enhanced to the detriment of another (mangroves or unpierced shores). This clearly points to the complexities and contradictions inherent in environmental engineering and disaster management. It also highlights the micropolitics of loss and gain.

In this case, the weak and fragile embankment was a source of income only to Aditya, his gang, and their families. The "resilience trade-off," consequential for all living in the shadow of this particular stretch, held a tangible economic benefit only for those who had managed to get a shift.

This was not the only take on mangroves and their proactive role in coastal protection assemblages. Many islanders look at them as producers of new land. On exactly these terms, mangroves have been planted along some of Sagar's edge. Staying clear of a language of sea level rise most of the time, politicians pitched them as means of counteracting coastal erosion by instigation accretion. Mangrove roots—one Gram Panchayat leader told me—make new mud. Sediments, he explained, settle at their tangled roots, and new land appears over the stretch of a few years under favorable conditions. He was not alone in this assessment. Hydrologists echo these claims, which in turn find their way into strategy papers or humanitarian efforts seeking to proactively counter the threat of sea level rise and submerging coasts (see, for instance, Spalding et al. 2014). Nature conservation and shoreline protection go hand in hand.

On this island, plagued by ever-tightening scarcity of space to settle, mangroves were valuable for their capacity of churning out new land. From here, tensions ensued. Contrary to project designs and maps, which portray standing or newly planted mangrove patches as enduring presences, many islanders looked at them as intermediary formations that had to make way once their potential of building up land was fulfilled. Trees would be cut, and people would settle there as soon as possible, the leader told me, thereby clearly shortcutting the claim to perpetuity frequently built into afforestation drives, and demonstrating a view of forests that cherished them for the service of apparent land generation. In this view, their ability to churn out land seemed to matter more than the protection of shores. The role of mangroves as part of lively infrastructures thus remains deeply contested, and situated within political decisions on how to erect and to maintain infrastructure given financial constraints.

Considering the role of trees leads directly to a discussion of other, even less visible modes of assembling, and keeping in place, coastal protection.

Minor Infrastructures

Not very far from Jana's house, connected by a footpath on top the meandering outer embankment supposed to protect Ghoramara's eastern shore, Mondol used to own a greenhouse densely filled with pan creepers (*piper betle*). Each of the hundreds of plants steadily grew the leaves that give their name to and are an essential ingredient of pan, the mild stimulant and narcotic much loved across West Bengal. The market for fresh pan leaves never dries, Mondol tells me, answering the steady demand in labor and investments needed to keep the hothouse going with quick and comfortable returns. In fact, to Mondol, it was the only type of farming that provided an income at all.

And now it was threatened. Brackish waters had caught up with his greenhouse, submerging all that once stood between the wide estuary and his home-based business. It now sat immediately behind the outer embankment, on a stretch of high land.

Mondol knew that his greenhouse's days were numbered. He had seen and learned enough of those waters, of their rhythms, velocities, and hungers, to know that his greenhouse was bound to be hit by erosion and could very well go under this year. If he didn't act.

Like many other islanders, he had clung to hopes that the embankment standing between his key asset and the estuary would be renewed, and that, eventually, a more comprehensive approach would be implemented. He had sought to put pressure on local politicians to direct the administration's attention to the sorry state of the embankment he depended upon so badly. He told me of acts of handshaking and letter writing, of subservient pleading and frustrated demands, and of fatigued references to empty state coffers.

Nothing came of it. Only the monsoons were drawing closer. He decided to take luck into his own hands and chartered a small gang of workers. Over the course of a few days, they cut mud, hauling and heaving it atop the frail embankment, enforcing it around the perimeters of the greenhouse. The task was tedious, and futile. Once again, construction activities and abandonment were woven into one another. No one here thought that it would make a big difference; they merely hoped to buy time. Since the delicate creepers struggle in salty conditions, their demise was certain. But Mondol could harvest fresh leaves every day, and therefore any extension of operations meant a return.

Mondol's intervention was modest. His aim was to have his operation running just a bit longer, and make just a little more profit from a doomed plot. In so doing, he performed infrastructural labor, using his and wage laborers' bodies to add to frail and overburdened coastal protection (Ley 2018). He lent his muscles and private funds to bolster local undermaintained coastal protection. Unnoted and unacknowledged by responsible authorities, he invested in small-scale, tactic improvements in spite of knowing it to be a lost cause.

I understand Mondol's attempt to fortify this particular stretch of embankment as an example of "minor infrastructures." The notion of minor infrastructures captures hardly visible efforts of orchestrating bodies and matter to enhance the capacities of often-overburdened infrastructures. These efforts might appear rather inconsequential, merely providing relief or room to maneuver, delaying the inevitable. Building on Erin Manning's (2016) conceptualization of "minor gestures," I call attention to how specific kinds of gestures (involving muscles, energies, and matter), in themselves or through the traces they leave within a shifting waterscape, shape coastal protection in often-unacknowledged ways.

Scholars working on infrastructures tend to distinguish between major and minor infrastructures by reserving the first term for state-backed, large-scale attempts at provisioning. In such a view, the latter, minor infrastructures, capture "self-provisioning strategies" (Bresnihan and Hesse 2021, 12), essentially local alternatives to large-scale ones. With a nod to Deleuze, "minor" here stands to embrace marginality and the disruptive potential of other stories (Kemmer and Simone 2021). Considering the interest in unofficial attempts at providing alternatives under conditions of marginalization, I take major and minor modes of infrastructure as closely entangled. On contemporary coasts, at least in the Sundarbans, minor infrastructures are contained within, and enabled by, official coastal protection. It is a practice activated and best visible in moments when distributed disasters condense, or when acute crisis looms. Yet, it remains a challenge to account for these very practices, as they are fleeting, small-scale, and often hidden from view. Minor infrastructures call attention to *spectral dimensions,* the *afterlife of ruins,* and *everyday forms of care.* I address each of them by returning to islanders' struggles with unruly waters.

Mondol's endeavor to provide minor infrastructures is an illustration of *spectral dimensions.* Safeguarding the greenhouse was impossible, and its structure and plants were soon lost. Skirmishing with eroding grounds, heaving mud up only for it to be washed away sometime soon, spelled out how bodies and matter provide infrastructural effects for fickle moments in time, whose acknowledged futility renders them a half presence, somewhere between presence and absence. By this I do not mean all was in vain. Mondol managed to harvest and sell his precious leaves a while longer. But all this was made possible only by the privately provisioned extra fortification being steadily washed out, and therefore the mobilization of extra efforts that allowed an already forsaken greenhouse to persist as they were undone. Ephemeral and limited in scope, minor infrastructures demonstrate how islanders, far from being merely victims, navigate submergence and abandonment on their own terms, and how they seek to invigorate land-based economies even when all looks to be lost.

In addition to attempts at safeguarding private property, similar circumstances arise when islanders seek to cushion potential destructions during hazardous spring tides or surges. As noted in chapter 3, full moons are feared as times when spring tides reach highest and embankments are most likely to collapse. In such tense hours, people gather and toil to avoid breakdown and patch up collapsed defenses in order to minimize impacts. Bodies scramble to hold embankments. Some look out for fissures, others bring mud to patch up gaps. At worst, chest-deep in turbulent waters, facing nearly torn structures and struggling with mud and silt, they are themselves, for the moment, part of failing infrastructures. Unaccounted for. Rarely visible. Below officially sanctioned maintenance. A largely transient, yet practical attempt to avoid becoming displaced by an encroaching estuary holds consequences for the duration of the embankment and, therefore, for people next on the line still enjoying dryer lives (see Harms forthcoming).

Writing on Moushuni, an adjacent island also badly affected by coastal erosion, Danda (2007) reads such efforts as spontaneous collective actions. I concur with him on the ad hoc quality of these efforts. However, they are also regularized instances of dealing with abandonment that—in practice—subvert state claims to the control of public works (Sarkhel 2015).

Struggling with mud and water in moments of acute threat is, of course, in itself uneven. This refers to the specific sites where such interventions are focused. Minor infrastructures concentrate on outer ring embankments, whose collapse would mean the swamping of the homes, fields, and neighborhoods sitting in their shadow. This is what Mondol did, in a preemptive gesture. And this is what residents along threatened embankments continue to do during spring tides.

But not all embankments are fortified by minor infrastructures. This unevenness also reveals the considerations on what is worth struggling for and who can muster the means needed. While Mondol invested in his greenhouse, calculations would be different for rice fields. Among the vast majority of my interlocutors, rice was a subsistence crop (rather than a cash crop), hardly feeding the family for the entire year. Paddy fields, therefore, are much less valuable and less likely to attract the substantial investments Mondol needed to cough up for minor infrastructures for his greenhouse. This is aggravated by the long period between purchasing paddy seeds and harvesting, which makes rice a risky crop on any threatened plot, in contrast to the quick return of an already well-established pan greenhouse. Such considerations account for further unevenness of coastal protection, now through investments into or through the withholding of funds, attention, and care.

In the long run, Mondol's greenhouse left no trace. When I returned next time, it was all washed away. For some time in between, however, first the structure and, later, what remained of it must have been a hurdle to the waters attacking the island. In doing so, these remnants offered an example of the *afterlives of ruins* within coastal protection assemblages.

At some point, Mondol gave up, diverting his efforts elsewhere. I was not present when he took the decision, but have seen similar greenhouses abandoned to the waters. Just as with houses left for good, farmers extract all that can be used elsewhere, such as tarpaulin, creepers strong enough to plant elsewhere, and sometimes poles. All else is left to the waters, cluttering the area under liquefaction and hindering, however negligibly, the water's approach.

As structures collapse, so do the foundations provided for short-lived wave breakers. Once again, greenhouses are illustrative here. Since pan requires freshwater soil to flourish, islanders erect greenhouses on raised platforms that are hardened by treatment with earth, dung, and chemicals to ensure ideal grounds for the creepers. Once abandoned, these mounds cushion the waters before slowly disappearing.

The same happens with houses left to the waters as people leave for good, just like those dotting Jana's neighborhood, with which I started this chapter. They too provide for breakwaters as they get undermined, collapse, and disintegrate into bricks, tiles, or poles, cluttering the area and serving as a ghostly reminder of earlier residents.

Besides greenhouses and homes, similar can be said about normalized procedures of settling lands that evoke Bengali visions of the good life. Trees planted, roads laid, hand pumps assembled, or gravestones and memorial shrines erected, as is popular on these shores, all slow down erosion as anthropogenic interventions. Few of these are as sturdy as the occasional concrete-enforced sluice gate, frequently standing a number of years in the waters and as an outlier of already-receded land, their eventual decay simply a matter of time. Most are unintentional and fleeting in equal measure.

Minor infrastructures consist, finally, of *everyday forms of care* by islanders peopling eroding shores. Consider Zeynab. During the early days of my research, the girl lived in a ruined house in Botkhali together with her family of eight. When we first met, their house was in an area frequently flooded once the embankment collapsed, and when the ring embankment had been moved inland, the house stood in the abandoned zone susceptible to regular flooding (see figure 6.5). Now, both their house and the land it stood on are long gone, and the whole family has moved out permanently.

With meager resources hindering a move and with faint hopes of government intervention, they had clung on to their house for a while. They lived with regular seawater incursions into their neighborhood and their home. In those times, extra efforts were needed to enable modest forms of living in the flood zone. Being ordinary, and barely visible, they still intervened into the texture of the waterscape in ways that complemented the more easily discernible official and unofficial protection works. I am referring to mundane efforts of sustaining homes among floods.

The navigation of hazardous landscapes all too often presents women with an extra burden (Sultana 2010; Crow and Sultana 2002). In addition to sustaining families when husbands or fathers are frequently away on the labor migration circuits, and the troubles of gathering drinking water

Fig 6.5. Houses and footpaths beyond the ring embankment, Sagar Island. Photo by author.

in increasingly salty abandoned lands, everyday tasks of upholding these very houses prevail as usual. Gendered norms encumber girls and young women with the recurrent tasks of cleaning and plastering floors and houses.

For Zeynab, this demanded administering mud, dung, and water in daily rounds of house restoration. Through these mundane acts of repair, she contributed to the survival of the house. Her barely visible and normalized practices of care contributed to her family holding it out in the flood zone, and to the house remaining a hindrance to the tidal waters coming in with increasing fervor. Plastering the house joined a range of other barely visible activities of weathering floods, such as the unpaid and unauthorized laying of brick paths connecting houses to the next state-sanctioned road. These needed regular repairing as the waters were washing over them constantly, but they still offered a degree of defense. Otherwise there was the fortification of homesteads through the addition of mounds of mud to their outer perimeter, to make them withstand tides a little while longer. Again, these fortifications required regular maintenance—eyes checking for cracks, hands adding fresh mud—and were in the waters' way for as long as they stood. Just like proper embankments. If tempering and directing flows characterizes coastal protection infrastructures, the plastering of homes joins these assemblages, I suggest, by providing for minor infrastructures through mundane care.

Infrastructuring Coastal Archipelagoes

Thinking embankments as instruments of landscape transformation not only exposes their unevenness or their many constituents, from designs all the way to mundane practices of care. It also positions unevenly embanked islands, and their inhabitants, on a larger canvas. Unsurprisingly, islanders are very aware of that. Commenting on the failure of protecting Ghoramara from eroding waters, Nirmai Pradhan, stressed: "After Ghoramara is gone, the currents will hit Sagar, and then also Sagar will vanish." His words underlined the widespread notion that Ghoramara has been abandoned by the state, left to sink and sacrificed by an administration that routes funds and expertise into more-prestigious and less-futile projects.

His words also had a clear political message. He wanted the state to reconsider. Emphasizing the suffering by islanders forced to move along, and ultimately to move out for good, as their island was shrinking, was not enough. They had done that for decades to little avail. Instead, he called attention to the trouble others would face once their island was gone (see Harms forthcoming). The island might not have been a worthy candidate within the distribution of coastal protection efforts, but its position as an outlier for the other, more populous and famous island Sagar, might compel administrators to give it another thought.

Islanders on Sagar, in turn, sometimes used the same argument. Many times, I was told that Sagar need be protected, because were it not to persist—overrun by surges or sunken by erosions—the waters would soon be attacking the city of Kolkata. Again, a fairly small landmass came to be pitched as a protective outlier for a larger, more critical terrain, this time the target being India's third-largest city.

The future of both islands remains undecided. Even if Sagar appears to stand better chances, given its critical role within Hindu sacral geography, I have shown how this island is in itself fragmented by uneven protection efforts. When islanders depict Ghoramara or Sagar as protective outliers, they position comparatively large-scale, high-tech approaches as well as minor infrastructures as consequential for the future of the waterscape they are embedded in.

Chapter 7
Of Blame and Protection

Juddhistir Jana recedes into a moment of pause. It is getting late. We have spent the better part of another afternoon on his veranda. With the mattress rolled aside, and mosquito netting tucked into the bare roof, his bed doubles up as space to sit and talk. Driven by sweet tea speckled with milk powder, under the watchful eyes of Hindu deities looking down on us from calendar prints nailed to the bare brick wall, our conversation crosses back and forth between conventional interview and informal exchange. A man of age, he has many stories to share.

Juddhistir Jana's life is closely entangled with these vivacious waters. Several times he has had to wrestle the salty swamps for a place to dwell in. Now his sons are thriving by daring the sea, working full-time as fishermen on a big trawler. They are away most of the year, either fishing or resting and stocking up in the harbor. Visits home are always short. Every spring tide, Juddhistir hears the roaring waters rising along Colony Para's outer embankments. He is well aware that the *bhangon*, coastal erosion, he used to endure is now directly affecting people nearby. He knows that many islanders are troubled by whether the island can prevail against rising seas and shifting currents. On the veranda, the conversation moves from personal achievement over to a more general assessment of the waterscape. My question of what it was that makes the waters so treacherous, and their existence here on these shores so precarious, brought him to a halt.

He does not know, he says. People like me should know, he adds; that is, people who are educated and have seen the world. His reply was a common one. More often than not, islanders would turn the task of explanation back to me. A number of times, I have heard people insisting on not knowing. Uncomfortable silences take hold, fueled by a fetishization and racialization of erudition alongside the deeply engrained self-depreciation of village folks. This double move features widely in the annals of ethnographic fieldwork (e.g., Ortner 1995). It speaks of power relations between researchers and their "subjects," bound up in postcolonial histories. It may also highlight tacit knowledge, resisting attempts at exemplification, or simply a refusal to speak or share in the process of creating scientific knowledge. With Juddhistir Jana, such power relations were obviously at play, however hard I worked to push against them. Yet, it was not a refusal to speak nor the elusive nature of something tacit that made him pause and play the ball back into my court. It rather signaled, I argue, a profound uncertainty about the nature of the water and what its driving forces might be. Here a wide range of possible explanations, morphing in time much as the islands do, flick between lifelong experiences along capricious shores, religious belief, and human intervention.

Juddhistir's everyday life is not merely enveloped by brackish waters. It is shaped by divergent ways of engaging and addressing them, of making them matter within the everyday. At one point, exuding a sense of pride, he takes me to his private shrine. It is built on the outer perimeter of his garden, the gods watching out, or so it seems, across homesteads and Colony Para, and on toward the sea. His shrine features Ganga Ma, a goddess of fluid divinity. It points toward both the spiritual significance of the river and, arguably, the economic advantages garnered from catering to the never-ending stream of pilgrims. The money needed to build the shrine came from fishing, an activity that relates differently to the waters, by using machines and supplying regional markets, and one that lays bare the risks involved in dicing with this interface of river and sea. All that—the shrine, the veranda, the pride about his sons doing well—was in spite of resettlement after being ousted by a river gone rampant. But even if his life might appear to be soaked, he saw himself as firmly attached to land. Juddhistir Jana never doubted the continued existence of the island, and saw it as being protected from ultimate submergence—an act of futuring amid unruly waters and sinking lands. The pause and its context, in a word, signal multiple ways of relating that are driven, as I will suggest in this chapter, by profound uncertainties about who does the doing, and why.

With erosion ubiquitous, what animates the water remains up for debate. Water itself has an evasive quality. Its materiality is challenging to think with, as it is prone to leak, soak, muddy, or shift states (Strang 2004; Cunha 2018). But that was not the end of it. It rather turned out that the unruly, fleeting material entity "out there" is imbued in a web of meanings and attributions that transgress the neat divisions of society and environment, of immanence and transcendence, of now and then, or of matter and spirit. In this chapter, I engage with the diverse, complex patterns of water, its approaching, rising, and attacking. I join islanders in debating what it is that makes the waters push against the land, relentlessly advance, and voraciously engulf all there is. Has it always been like that? More importantly, is there a way to protect their island homes?

Where Juddhistir remained silent, many others would passionately mull over possible reasons. Flaring up in all kinds of moments, such conversations featured cascading uncertainties and loose threads. Across the preceding chapters, I have unpacked a firm awareness about the pace, rhythms, and scale of erosion on specific stretches of the coast. This chapter supplements my account of knowing, ascertaining, and anticipating by delineating deeper ontological insecurities about what or who might be behind coastal erosion.

This is no mere story of islanders left out of expertise sharing, but one of a profounder kind of uncertainty. It arises in having to deal with a vast waterscape that is hard to ascertain and rendered divine while being the object of mundane or spectacular human interventions. Obviously, islanders are very much aware of that conundrum. It is this very predicament that accounts, I suggest, for much of the discursive uncertainty that informs narrations of the disaster.

No theory alone can explain the curious behavior of currents and waves. Local accounts are nuanced yet ambivalent, oscillating between divinity, hydrology, engineering, and politics. Almost at will, these fields can be activated to make sense of *bhangon* in the Bengal Delta. But such analysis fails to grasp the ambiguities involved. It falls back upon canonical domains of Western modern thought—such as science, religion, or economy—whose normative leanings and rigid distinctions help little to tackle modern problems (Latour 1993; Asad 1993). Currents and waves shed light on the many textures and drivers of the water, its many natures, and their entanglements.

Explanations are diverse. They differ in tone and content, from place to place, from person to person. Even when blame is firmly attributed to one source, a set of further dimensions, subtexts, and metaphorical relations belie any straightforwardness and complicate constructions of causality. If anything, these entanglements demonstrate the limited value of a conceptual triad of materiality, divinity, and humanity in attempts to disentangle notions of causality of disasters. I will therefore not guide my analysis by moving through a sequence of domains, but rather move back and identify mutualities and entanglements as I go along. Typical social science categorizations, such as gender, class, and place, or of science and religion can conceal as much as they reveal. Instead, these narrations lay bare fickle patterns, code shifting, and ambiguity.

If my data forecloses the mobilization of clear-cut labels, and unearths ambivalences and quick shifts in making sense of the water's hostility or its limits, how do I do justice to this within a chapter of an academic monograph? How can I make this resonate within the affordances and demands of a text based on making an argument? In other words, how can I contain this multiplicity in a text banking on a sense of unambiguity? My account will remain loose, the threads not firmly entangled. I do not aim at imposing some kind of order. I provide provisional takes. This might frustrate expectations invested in social science analyses. Yet, doing so means staying truthful to my friends and interlocutors who channel such ambiguities, shifting between codes. One of the challenges here consists of the politics of knowledge and authority, when Juddhistir Jana, for instance appeared to restrain his voice.

I argue that islanders' uncertainty about how to attribute blame partly flows from the texture of the disaster. The process of rendering disastrous events meaningful is in itself distributed unevenly across time and space. Much of this turns insights from typical disasters on their heads. If conventional disasters seem to strike out of the blue, in most cases they become the subjects of rather uniform explanations through short-lived but intense debates (Neiman 2002). Coastal erosion differs from this once again. It is never out of the blue, but subject, I argue, to kaleidoscopic visions of how it comes about and who controls it. Coastal erosion appears as a well-contoured, steady presence, and there is not one moment nor one event demanding collective explanation. In coastal erosion, aftermath and prognoses are distributed themselves, in knotty terms, inviting uncertainties to take hold. Considering waves and currents, and what drives them, simultaneously shows how distributed disasters refract environmental and political relations in nonlinear ways. This is relevant, I propose, not only for understanding lives engulfed by coastal erosion. It also indicates the conceptual and practical troubles that come with environmental degradations that are at once low-key and catastrophic, and, more broadly, open-ended yet knotty. In other words, uncertainty at the waterfront has a lot to teach us on the challenges of coming to terms with the Anthropocene. I begin by exploring islanders' take on fluid divinity.

The Other Ganga

Consider a map of Sagar Islands, or the Sundarbans at large. True to cartographical traditions, maps rely on the vectors of sea and river. Letters and color schemes situate the sea as the force washing against the southernmost shores, while everywhere else rivers flow. Islanders' vernacular conversations provide a different picture. To them, the waters around the island in all directions were river. Standing on the southern beach, looking out at what maps consider the Bay of Bengal, teams of fishermen assess that day's character of the river to decide when best to catch. That is, the waters were for most purposes qualified neither as marine nor a mix, but simply as riverine. To them, they appeared the body of the river, unfettered and undiluted, pushing out into the sea and engulfing those tiny islands. This resonates with mainstream Hindu accounts referring to the region as a whole, and Sagar in particular, as the confluence (*sangam*) of Ganga and the sea.

The sea existed, of course. As a distinct environment, a source of hazard, a space of danger, and a realm to mine. Contrary to a cartographer's visions, however, the sea only sets in after one has left the shore and is journeying aboard a trawler southward for several hours. The shift is gradual and subject to tides and seasons, indicating an acute awareness of estuarine fluidities pushing deep into the ocean. Waters of a brownish hue, whose waves were tamed by the pull and weight of sediments, belonged to the river, while more dynamic, lighter-colored waters rocked by fiercer waves—free of sediments—signified the beginning of the sea. In this view, the sea never reached beaches and embankments, its existence cloudy to all who never went fishing.

Seeing the waters playing around the island as river, and river alone, is critical for understanding local ways of dealing with *bhangon* and other water-related disasters, for it classifies all waters, and their threats, as the activity of a singular river, Ganga. As Ganga, the waters washing around the island appeared not as a symbol or aspect or index of the goddess, but as her true being. Not as an abode, but her flesh and bone.

While the perception of all surrounding waters as the river, and indeed riverine goddess, was shared by the majority of islanders and pilgrims, practical engagement with the estuarine waters differed significantly. By this I am not primarily referring to islanders using the river as a source of income or livelihood, hauling in fish or catering to devout pilgrims. I am pointing to another way of relating to the divinity, of engaging her in rituals and everyday speech that are in tension with how Ganga is thought of and prayed to in most of South Asia, and the world.

In the pilgrimage center, temples and shops blare religious chants praising Ganga or the virtue of this *tirtha,* this pilgrimage site, or stories told by professional orators narrating mythical versions of Ganga's descent. The air is thick with devotion to a generous goddess, whose very body is—or so the narrators insist—relentlessly washing ashore in small waves around them. Likewise, on rikshaws pulled by islanders, families from across the country, but mostly from the North Indian Gangetic plains, cheerfully boast that they have finally made it down here to immerse themselves into Ganga's purifying body. For many, they are fulfilling a lifelong wish to bathe here and pay homage, hoping for the river to purify sins and provide blessings at this auspicious place.

To islanders, such encounters with devotional media or anxious pilgrims clash with their own versions of what the waters are. Newly arrived customers seek to get close and, more importantly, consider the waters to be purifying and auspicious. For local rikshaw drivers or stall owner, perceptions are more ambiguous. Many had seen their homes dismantled by waves, and therefore desire to stay clear of the waters as much as possible. Yet roles as tour guides to pilgrims were still attractive to many islanders, and in those roles they would rehearse talking points taken from mainstream accounts as they transported them and, thereby, reiteratively emphasized (also to themselves) visions of good and overly auspicious waters.

Kelley Alley engages with a similar problem (Alley 1994, 2003). Some thousand kilometers upstream, in the pilgrimage center of Varanasi, people struggle with the mounting pollution of the sacred river that is both a lifeline to the city and the object of veneration. Alley argues that residents and visitors distinguish between pure divinity and material pollution floating along. In the river, both are entangled yet distinct, thus allowing purification amid the garbage, sewage, and bacteria not only flowing along but literally making up the river. In light of this bifurcation, there is grace in the stink. The pure Ganga is separate from the litter, set aside from the endless streams of waste bobbing around.

People in Colony Para do not draw such distinctions. Divinity also shows her grace and playfulness, as I will show, through roughened-up waves and augmented currents. In place of rescuing this benevolent mother from the hold of worldly blights, islanders entertain visions of the river that accommodate voracity and haphazardness. Listening to them, getting attuned alongside them to the water's ways, exposes a different vision of Ganga. Not one resuscitated within troublesome water and debris, but one that differs from the Ganga that most Hindus across the subcontinent and beyond bow to. I call this figuration "the other Ganga."

Considering ideas and practices, I tease out disparities between the way Ganga is thought of by mainstream Hindu philosophy and island-ers respectively. But to frame my narrative by two opposing Gangas—one mainstream, one othered—runs the risk of obfuscating what is at the heart of my analysis: uncertainties and the simultaneity of either take. In other words, the two Gangas overlap and form close circuits, and not only in crisis talk.

The other Ganga hardly surfaces in direct speech. Perhaps this is too risky. It certainly is problematic within the orbit of the sacral center that serves as a steady source of iterations of "Ganga the pure," and of income and status gleaned from her. It contravenes the North Indian–influenced mainstream, a steady force of influence and power. The other Ganga is mentioned more surreptitiously, if rarely in direct speech, and she crops up in certain quotidian practices.

I understand this other Ganga to be rarely explicit and more of a substratum, a hidden—at times not particularly well-articulated—take on the qualities and the form of the goddess, another vision and practice of proper conduct. Put differently, it was another goddess residing in, being, and acting as the river. This substratum became evident in diverse situa-tions, always retaining a hidden and profoundly ambivalent quality. It was expressed in everyday forms of conduct, in speech patterns, in the orienta-tion of festivities. Yet, it always remained clouded by ambiguity and vague-ness, undergirded, I argue, by a sense of uncertainty and apprehension.

Grace

Unruly and erratic, yet life-sustaining, rivers frequently attract notions of playfulness. Where the holy river meets the sea, this inherent playfulness (*khela*) coincides with grace (*mohima*) and duty or law (*dharma*). Islanders widely employ all three notions when reflecting on what makes the river act as it does, when debating the forces and rea-sons behind these unremitting waves and currents. Both inform elabora-tions on the thorny issue of theodicy: the question of why deities considered fundamentally benevolent inflict harm.

Islanders would tell me that erosions or, more precisely, the onslaughts of waves and currents are the river's grace, *nadi-r mohima*. Alternatively, I would learn that the river behaves in the way it does by way of its *mohima*. Colloquially, *mohima* stands for "divine grace," the fundamentally incomprehensible doings of divinity that collide with human demands or morals, follow their own logic, and are beyond the command of humans. *Mohima* reverberates across all major religious denominations present in coastal Bengal. It may very well be seen as yet another vector of syncretism characteristic of religious life on the delta frontier, which is arguably unraveling at the tug-of-war between the cultural politics of Hindu nationalism and Islamic reform. I choose to bypass an exploration of how these trends play out here. I am more interested in how the grace relates to the haphazardness of the river, rendering it the outcome of agency that is at once material and divine.

I would also often hear islanders referring to *dharma*. Taking land and futures, in this view, was a result of the river following her dharma. Within these contexts, dharma refers to the situated and personalized obligation of acting specifically. Across South Asia, moral codes and duties are not general, but rather the function of one's standing within hierarchical relations. Importantly, the usage of both notions when addressing erosion, or other ways the river impacts the lives of islanders, underwrites an apotheotic take on the water. Persons, including deities, hold dharma; and only gods, a singular God, or godlike people can make their grace felt.

On a more fundamental level, grace, dharma, and play introduce the river as a person. This is not specific to South Asia. Across the world, rivers tend to be anthropomorphized. Nor is the veneration of rivers as divine limited to South Asia, even if it continues to serve as a frame for thinking Hinduism and Indian nation building in general. That being said, it matters as what kind of person islanders venerate Ganga, how they relate to her, and how she figures in quotidian practices of interpellation or in public understandings of watery divinity. Among islanders, the river's grace, playfulness, and dharma depict a being that is more erratic than nourishing, one that unleashes suffering through its grace embodied in waves and currents, playfully giving and taking as she wishes. If islanders' articulations hold true, they are fundamentally incapable of understanding the rationale of the dharma that compels them to seek the protection of actors capable of defending them from the whims of the river. This alludes to a tension between the river and the communities that line its banks, while also underwriting notions of it as an erratic and hazardous brute more than a nourishing or bountiful goddess.

Digestion

The colloquial portrayal of the river as a person—and a troublesome one at that—also relates to the waters through a language of digestion and eating. If waves and currents do the breaking (*dheu bhengeche, srot bhengeche*), the river swallows land and futures. Thus, Jagganath Maiti (2008, 11), the locally renowned historian of Sagar and adjoining islets, notes in his accounts of island life that after living at ease for some time, at one point for "the villagers [of Ghoramara's Khasimara] difficult times set in. Slowly areas began disappearing in the awful mouth of the Hugli River and were to remain to stay in its stomach." Colloquial Bengali certainly is replete with phrases and compound verbs referring to ingestion. Smoking translates into, literally, the eating of cigarettes; getting beaten up translates into eating blows. Similarly, the stomach serves as an indicator of stress, anxiety, or ease in ways that render social and psychological conditions and metabolic activity mutually susceptible to one another (Ecks 2013). In both ways, digestion appears as a human characteristic, and its particularities are sites of articulating and crafting personhood.

Note that Maiti uses the name Hugli in this quote. This may have to do with accuracy, relying on the unambiguous name under which this tributary is known, or with piety, reserving the label Ganga for more auspicious interactions. Most fellow islanders, however, freely and persistently used the label Ganga, or simply "river," when referring to the onslaughts of salty waters in the form of waves and currents. Frequently, they would state that they were landless because their holdings had literally gone into the river (*ganga chale gecche*). At other times, they would refer to land being eaten up by the river Ganga (for instance, *ganga kheyeche, khete-khete ganga nadi chale gecche*) or disappearing by way of being eaten (*khete khete chale gechhe*).

Echoing Maiti's language of oral cavities, islanders would also consistently refer to the digestive apparatus—lands had disappeared in Ganga's mouth (*Ganga mukher modhye adrsya haye gache*) or now sat in her stomach (*pete thakbe*). Nimai Pradhan, an islander resettled in Colony Para, explained in what appears to be an idiomatic statement that "it was the Ganga that has eaten it [the land] up and now it rests in her stomach."

To be sure, it is not only Ganga who eats and digests. Other oral cavities feature as well. Sometimes people would refer to coastal erosion as the reality of literally being "in the mouth of erosion" (*bhangoner mukhe*). As a whole, Sagar and Ghoramara islands sit in the river's mouth, as deltas frequently are called. And while this arguably refers to deltas functioning as an orifice through which rivers enter into the sea, it also plays on the vast metabolic potential of the waterscape, a site of churning, absorbing, and discharging.

In these iterations, the river appears as a person who is more erratic and troublesome than nourishing or purifying. She troubles by being voracious, eating relentlessly into the land, the very foundation of survival and well-being on this rural frontier.

In the Way of Capital

One day, I spoke to Malini Devi as she took a break from her chores. It was late morning, her husband and sons off chasing wages. She had a little time on her hands, and we eased into conversation as she sat in her family's cycle rikshaw parked in a shady spot in the porch of her house. As the discussion drifted toward *bhangon,* and how life used to be before they moved to Colony Para, she became tense. "There, where now the ships are passing, once my home stood," she erupted. It was said with anger, fueled by the injustice she saw at work here. Ships now traversed her former homestead, as if nothing had happened, using what used to be family grounds as a shortcut on their way to or from the port.

But this was not merely a temporal sequencing, as if ships had simply taken over a space created by usurping tides. To her it was an account of causality, as she made clear. What had made the waters break her land and her former house was, she explained, a matter of powerful currents. And these currents, she went on, her voice wavering, were produced by container ships as they passed by what used to be her island on their way to the port. Over the course of months and years, I heard similar renditions countless times. Virtually idiomatic, such accounts never failed to express sorrow and anger at disinterested people steering monstrous ships, oblivious to the toll their actions were having on the lives they sailed past.

Just like Devi's son, many saw the recent upsurge in erosion as stemming from the subsurface walls that port authorities had erected close to the main entrance to Haldia Port a few decades ago. These interventions, they agreed, had redirected the troublesome outgoing tides away from the port and directly toward Ghoramara, so they were now hitting its northwestern flanks with full force. Outgoing tides had already been causing much misery, dragging chunks of land back with them toward sea. The walls increased this tremendously. In this view, currents churning the waterscape are conundrums of tidal flows and seasonal vagaries, but they are crucially tweaked by human interventions.

Islanders are acutely aware that attempts at making accurate claims are futile. Trends can be gauged and interventions intensifying currents pinpointed. Here, the fluid body of divinity becomes protuberant, and fluvial playfulness gets tweaked, with troublesome consequences. Attempts at checking the port's silting may further mobilize machines of destruction and increase the voracity of an already troublesome river.

Water as a public good, and the relations resulting from this, have long occupied the gaze of anthropologists. Infrastructure development refracts such relations. Much of the literature concerns either the maintenance of communities through water networks, such as in Geertz's (1972) classic account of Balinese irrigation systems, or the socially detrimental effects of rerouted or upended flows, such as when reduced discharge proves a burden for downstream communities (Rahim, Mukhopadhyay, and Sarkar 2008). Examining currents on Ghoramara, however, reveals a different picture. Here we see relations sustained by water whose excess quantity and directedness harbor devastation. In fact, the vast amount of sediments the estuary is carrying had been a problem as long as the port of Calcutta existed. Port engineers were engaged in endless debates as to how to secure safe passage for cargo ships. The issue became more pressing as ships continuously grew in size. Dredging has been the most convenient solution for centuries, with a dedicated armada of dredgers moving up and down the channel razing the bed within tight parameters for other vessels to pass (N. Mukherjee 1968). Another attempt has been to erect a new port closer to the mouth of the river, at Haldia, thereby reducing the enormous expenses levied on the treasury. But shortly after the opening of the new port, the old problem returned. In fact, the intervention again disturbed hydrological processes, arguably aggravating sedimentation. Eventually port authorities reacted, and built subsurface walls beyond the channels, which kept much of the tidal activities out of reach of the port channel. While this mostly alleviated the port's tribulations, the walls redirected the vast majority of the tidal outflux waters away from the channel and toward Sagar on the opposite shore. It is unclear if the responsible engineers underestimated the devastating effect this redirection of flow could have on the opposite bank, or if it was irrelevant to them. In any case, observers agree that the walls worsened erosion profoundly. Newspaper articles voiced concerns (see, e.g., A. Ray 2009), and islanders also noted the changes. Walls and banks are tied to one another in what Johnson (2019) calls "spectral flows," long-distance connections that are barely visible and of a ghostlike quality, yet remain deeply problematic for riverside dwellers. If the spectral qualities of waves and currents made for long, sometimes fraught discussions, they flagged a sense of being overpowered, of being at the receiving end of unjust relations sustained through water, and of being at the mercy of larger-than-life agents wielding their power—knowingly or unknowingly—through estuarine waters.

While all the waters encompassing the islands were simultaneously river waters and the body of the goddess, what made them move was not divine activity alone. If currents and waves accounted for islanders' miseries, gobbling up homesteads and undermining futures, they were not the root of all troubles. The vicissitudes of estuarine flows appeared as embodiments of unnaturalized hydraulics, as conflations of water, divinity, and technological interventions. Muddy flows meet muddied explanations. Hydrology turns out to be a fraught and messy terrain, where politics, technology, and ritual intertwine under conditions of uncertainty.

This is not purely a conceptual matter. Attributing blame involves deliberations of possible futures as well as of responsibility and justice. This was the case when Devi blamed towering container ships. It was the case when people like her son blamed port authorities for carelessly constructing walls. Both diesel-fueled steel giants passing the islands and currents silently steered toward them signaled the carelessness of aloof authorities intervening in the waterscape as they pleased. Ocean-borne traffic and petrochemical industries were thriving at the expense of, and in utter disregard for, rural islands. Suffering for the benefit of industrial development was in many places counted as sacrifice for the nation (Parry 2008; Maldonado 2018), and islanders saw themselves at the receiving end of "uncare" (A. Gupta 2012) that was not even acknowledged. Ships and subsurface walls condense deeper layers of injustice, illuminating a sense of denial and displacement of the rights of people standing in the way of capital. It amounted to being left bruised by the machinery of the developmental state operating at full steam, or so it appeared, and without concerning itself with the detrimental effects its activity would have on other lives close by. Debating waves and currents here serves as a critical commentary on the relations sprouting from these waters. The advent of climate change and sea level rise further complicate this conundrum.

Bringing climate change into the equation requires care. There is a "working misunderstanding" (Sahlins 1987) at play. If asked directly about climate change, islanders would frequently agree that it exists, of course, before immediately detailing the change of the seasons.

Rudiak-Gould (2012) notes similar forms of what he calls "promiscuous corroboration" in his study of the way climate change is communicated between islanders, researchers, and journalists in the Pacific Islands.

More interestingly, in local conversations, the rhetoric surrounding climate revolved around anticipating surges and floods. The run-up to the 2009 United Nations Climate Change Conference in Copenhagen, which coincided with the beginning of my work on Sagar, seemed decisive for this development. Pollution and excess emissions from parts of the globe undergoing industrialization would translate, I learned in conversations on Sagar and Ghoramara, into tsunamis threatening these low-lying shores, or so my interlocutors had learned on the radio. This version beholds sea level rise as an issue of disastrous surges, and much less one of a permanent near-universal change in water levels. It obfuscates the distributed effects of shifting water levels for the benefit of dystopian visions of catastrophist proportions. On another level, there is a causal connection between industrial development and its life-squashing consequences, often unfolding belatedly. Over the years, the imagery of a tsunami unleashed by polluting emissions has lost much of its grip on islanders, and anthropogenic climate change continues to be overshadowed by the combined troubles of passing ships, infrastructural walls, and playful rivers. But a tsunami triggered by climate change—however figurative it may be—similarly articulates a sense of being in capital's way. Boat propellers, vessels, and walls all epitomize projects of industrialized, global flows of goods or resources that not only leave rural islanders stranded and with little trickle-down benefits, but literally undo the material underpinning of their existence.

Everyday encounters with waves and currents play against any conceptual takes on their character. Before exploring the imagery of protection, I turn to mundane stances and habits and their take on a roughened up, uncertain river.

Desiring Distance

For many islanders, their Ganga was one to avoid. Most of those I know hardly spend time on the waterfront, unless there on some kind of business. Fishing in all its iterations takes place on the riverfront, including catching *meen*, setting sail on trawlers, hawking catches, or drying them. Seeking income from the pilgrimage traffic also frequently requires taking to the shore, where residents sell ritual paraphernalia that pilgrims use to serve their Ganga. Beyond these activities, the waterfront rarely figures as a place to hang around or enjoy. This stands in contrast to pilgrims and tourists, who take to the shores for rest, enjoy sunsets, and engage in elaborate selfie safaris. It also diverges from the way riverbanks figure as vivid social sites to relax, socialize, or wash in the pilgrimage towns lining Ganga, and her smaller sisters, all across Northern India (Haberman 2006; Alley 2003; Drew 2017; Doron, Barz, and Nelson 2015).

Similarly, ritual bathing in liquid divinity was left to pilgrims. Taking a break from his tasks in the high season of drying fish, Pashupatti Giri told me jestingly that fishermen are in the water all the time and therefore do not need to take ritualized baths. The joke plays on taking being in the water as the key element of ritual ablutions, whereas in the worlds of popular Hinduism the latter consist ideally of orchestrated bodily postures, spoken words, and affective states. Even pious Anil Patro, who was in charge, for a time, of looking after one of Colony Para's public temples, restricted the bathing routine that was required as a precondition of serving the deities to bathing in his pond. Yet, he still joined the masses during the pilgrimage fair in bathing at times considered most auspicious, even when the vast majority of residents did not.

To pilgrims, on the other hand, bathing marks the high point of ritual. It promises, as noted, the ablution of sins in the purifying waters of a benevolent river. The skirmishes that occur in making it to the shore, and into the waters, at astrologically defined, auspicious moments during the festival evidence this. Here lies an unmistakable difference in ritual logics and encounters with Ganga. Where some groups plan whole trips and journeys around immersion in these sacred waters—and jostle with one another to do so—those who live close by actively avoid them.

Anxiety surrounding rivers is not specific to Sagar Island or Ganga in her lowest reaches. It can be found all across the Bengal delta. Jalais tells of mistrust toward rivers among islanders living in the heart of the Sundarbans, close to the mangrove swamps, some eighty kilometers from the coastal fringes of the archipelago (2010b). One of the major tributaries of the Ganga feeding the area where Jalais worked is unambiguously called Matla, which translates from Bengali into "intoxicated" or "drunk." Elsewhere, there is Damodaran, literally meaning "sorrow," which empties into the Hugli not far upstream from Kolkata.

Acts of avoidance, distancing, or worship convey a sense of being alert to eluding punishment, and resonate with the way what are known as hot goddesses or village mothers are viewed in rural Bengal. In his classic account of ritual practice in rural Bengal, Ralph Nicholas argues that the hot goddesses Manasa and Sitala are related to particular environmental

hazards (1981, 2003). Combining historic and ethnographic data per-
taining to the Midnapur district, just across the part of the Ganga River
that most of my interlocutors call their ancestral home (*bari*), he argues
that the rituals surrounding these goddesses arose as an answer to mortality from fevers and snake bites, respectively, and work as a means of
coming to terms with these troubling experiences. While I find the func-
tionalist hinge of Nicholas's argument problematic, his take is illuminating for thinking water relations implicated in coastal erosion. It seems
safe to say that Manasa and Sitala draw on, and reiteratively inscribe, a
mutual identification of a particular deity with a particular form of
embodied suffering. The snake goddess Manasa is closely related to the
presence of deadly snakes, and the experiences locals have with poisonous ones. Sitala is closely related to fevers, spelling out mass mortality
before the spread of modern vaccines, with dear ones lost to disease or,
if all went well, only marked by clearly visible scars (Dimock 1976; Greenough 2003). While, as Ferrari (2007; 2010) argues, sick bodies are
seen as the embodiment of goddesses themselves, their visits remain an
existential threat to well-being. The same applies, I argue, to the other
Ganga. Islanders relate to her through appeasement and distancing, and
not by purification or immersion. Interestingly, Nicholas makes no mention of ritual activity toward Ganga, indicating that the inclusion of
Ganga among islanders is a more recent phenomenon. On contemporary
Sagar, temple setups also underline this proximity very well. In private
shrines, such as the one Juddhistir Jana built and operated adjacent to
his house, an anthropomorphic Ganga is situated alongside the hot goddesses Manasa and Sitala, and included into the same worship routines
(see figure 7.1).

Branding it "the other Ganga" nods toward the profound unease
about the nature of divinity. Islanders' experiences with a goddess gone
mad and threatening what is most precious to farmers in an overpopu-
lated region—land and homes—underpin the desire to hold her at bay.
The other Ganga's activities are referred to in terms of monstrosity and
calamity.

Pacifying and Rejuvenating the River

Besides in speech and embodied anxieties, the other Ganga gains
traction in practices of worshiping. Islanders did not simply leave Ganga
to the pilgrims. Far from it. Ganga was regularly and intensively worshiped. But intentions differed, as did the language and the materials used

Fig 7.1. Ganga, Sitala, and Manasa, private shrine, Colony Para, Sagar Island. Photo by author.

in rituals. In this section, I explore islanders' worship of the other Ganga. I uncover two modes of relating to her. One is concerned with safe passage, the other with rejuvenation, yet both are intertwined in festivities as well as in everyday ritual conduct.

Safe passage is a chief motif, of course, among fishermen taking to the river and to the sea. Upon setting sail, they would turn to Ganga and pray for safe return. On their way across the estuary, travelers onboard ferryboats would also turn to Ganga Ma, asking for safe arrival. Such quotidian practices put forward a take on the goddess shrouded in risk. Juddhistir Jana, with whom I opened this chapter, is a fitting example here. In his private shrine and beyond, he primarily prays to Ganga for his sons to return safely from their fishing tours. After all, daring the waters harbors great risks. Engines are prone to fail, storms can hit anytime, and regularly overloaded boats subside all too often in the treacherous waters of the estuary. Fishermen regularly disappear. Hardly making news, anxieties fester in virtually every household. The fabled "tiger widows" of the Sundarbans are joined by what I think of as "water widows."

Ganga here emerges as one who is better appeased in order for the worshipper to be left alone and to not return scathed from an encounter with her. Here, she is a volatile actor demanding worship and a force to be wary of, sending punishment to those who would not bow—very much like the other hot goddesses dominating the landscape.

Besides safe passage, fish workers also turn to Ganga to ask and thank for catches. This manifests in daily rituals as well as a yearly festival held in her honor. The festival, Ganga Puja, pleads with the goddess to pacify the river, regenerate sustenance, and enable the material grounds for future prosperity. This festival is not to be confused with and is entirely set apart from the much larger, more famous Gangasagar Mela. In fact, the differences are important. Ganga Puja's focus on mercy and pacification arguably serves as the most striking and easily discernible expression of that other Ganga and her resemblance of popular goddess worship across Bengal.

Both festivals take place on the vast mela ground stretching between the beach and Gangasagar's main temple. Both combine amusement with ritual and sacrifice, featuring the sale of trinkets, blaring sound systems, and a funfair. Differences between the festivities, and the visions of divinity they embody, are engrained in the physical and organizational setup.

The timings of the respective festivals are revealing. Gangasagar Mela occurs around Makar Sankranti, which is on January 14 every year. As with most Bengali Hindu festivities, Ganga Puja follows the lunar calendar and is celebrated during the full moon of March or April. This timing underwrites the specific significance full moons carry along the coast, as combined moments of festivity and endangerment (see chapter 3). It situates this festivity firmly among those for so-called village mothers or hot goddesses, all celebrated for two to three days during a specific full moon (Nicholas 2003; McDermott 2013). Upon my inquiring why Ganga Pura happens at this time, my interlocutors provide me with two sets of reasons. Often, I am told this is simply the way it is done. Across Bengal, Hindus worship Ganga in their homes on these days, but usually not through public festivities. Conversely, people in Colony Para informed me that they are busy before, during, and after Makar Sakranti with catering for pilgrims, which makes it impossible to have their own rituals during that time of the year.[1]

Another key difference is that those organizing and performing the Ganga Puja bypass the main temple entirely. A committee of villagers plans and arranges all details, including the collection of funds within

---

1. Danda (2007, 122) notes that fishermen on nearby Moushuni have their Ganga Puja on Makar Sankranti in January.

the village. Similarly, the priests and musicians as well as entertainers booked for the festivities are pooled from the island itself. Ganga Puja is hosted in temporary structures assembled from wood, bamboo, and cloth. Electricity is provided by generators and water fetched from public pumps, bypassing the main temple's infrastructure. By contrast, the North Indian monastic order controlling Gangasagar's main temple, the Ramanandi Sampradaya (Veer 1988), arranges for the entire worship to take place in the main temple and without pooling resources locally. The mutual noninvolvement between main temple staff and village festivities accentuates islanders' detachment from the main temple and widespread mistrust toward its temple authorities. This in turn only seems to widen the gulf between the Ganga popular in mainstream Hindu folklore and the one I am speaking of here.

To islanders, the festival is also a time to celebrate abundance. Conversations hitting the subject were ripe with allusions of sumptuous feasts and piles of delicate fish. Within Hindu ritual, such foods are prasad. That is, they are both a sacrifice and a way of receiving blessing. Eating sacrificed foods collectively becomes a mode of receiving the blessing of the respective deity by either sharing a meal with her or by eating her leftovers. Feasting, in other words, is both a way of articulating a relationship with a deity and fulfilling perceived duties or demands and a way of ingesting the fruits of that work (Michaels 2006). If mainstream North Indian articulations of Hinduism are increasingly informed by visions of vegetarian *prasads,* notably also with regard to their Ganga, islanders on Sagar indulge in fish.

Rural Bengalis engage with such obligations largely in terms of worldly matters. Echoing Juddhistir's reasoning for worshiping Ganga in his private shrine, Maitri Devi, resident of Colony Para, told me that the puja seeks Gang Ma's benevolence toward those who fish in her waters. Celebrating Ganga by sacrificing fish—among other practices— is a means of securing passage, of making her abstain from capsizing boats and drowning loved ones in the open waters. At the same time, abundance and indulgence also engage with mutual forms of regeneration. Islanders told me that they perform the ritual to restore force and ensure future bounties. Fish donated, sacrificed, and feasted on is a means of giving back to the river, of enabling its rejuvenation.

In her study of dock workers up the river, Laura Bear also notes intricately interwoven layers of rejuvenation and protection. Exploring festivities held in honor of Ganga Ma, she notes (2012, 194) that "the worship of Ma Ganga particularly underlines a mutual labor of renewal between worshippers and the goddess that fills the material world with life. . . . They said it [the puja] showed their 'respect' for the river as a dangerous force that had to be mastered by marine skills, just as Siva had controlled the Ganga's force as she fell to earth."

Both modes—safety at sea and rejuvenation—allude to a Ganga that needs to be engaged with in order for one to secure some degree of benevolence and mutual sustenance, and thus she is a potentially erratic and dangerous deity dominating the waterscape that is to be reckoned with. Here a paradox comes into view. Through the Ganga Puja, people relate to the divine river as a dangerous mother, as a hot and erratic goddess that holds many threats. She may withhold catches or drown people and their boats at her whim. Alongside these real threats, the other dangers that islanders very often classify as the real disaster—like coastal erosion—remain unaddressed.

So far, I have argued that the water has many faces. It is a deity, an agent of purification, the body of god. It is a space churned by capital, a realm of fish waiting to be caught, a treacherous domain threatening to plunge seafaring men into its depths, a source of severe storms, and a waterbody scathing shores as it shifts and transforms, and lately also rises. It is all of this, and much of it at the same time. I have explored an uncertainty about how to address and what to do with waters that are at once divine, nourishing, engineered, and devastating. But even if ritual or mundane attempts to quell Ganga appear to be lacking, islanders do not perceive themselves as hapless or at the mercy of the waves. In the next section, I look at how protection is figured. Unsurprisingly, these figurations also draw on divergent ontological orders.

Hidden in Scriptures

Despite recent storms and the menacing presence of *bhangon*, most people living in Colony Para felt safe. In an extended conversation blending group discussion, formal interview, and relaxed chatter, Subrata Ray, a priest, captured the fundamentals of this sense of security very eloquently. "When after Aila struck," he exclaimed, "the helpers sent by the government arrived, they saw that on Sagar only one house had been destroyed, but that the neighboring islands had become like cremation grounds—this is due to Kapil Muni."

Subrata's hyperbole cannot hide the fact, nor did he intend it to, that Sagar sustained damage from Aila and ensuing cyclones that hit land in 2019 and 2020. But in the particularities of their arrival, islanders felt protected. And this sense of protection seemed to buttress futures for the island even amid coastal erosion and rising seas.

A number of islanders saw Kapil Muni's protective hand at work. He had allegedly ensured that none of these cyclones rolled over the coast during either the rainy season or a spring tide, when water levels are significantly higher. By tweaking with the storm's timing, islanders insisted, Kapil Muni ensured minimal damages. Many saw the divine yogi as having rerouted these storms away from the island's center, bending their trajectory eastward. Islanders combined evidence from the southeastern edges of the island, around Botkhali, where recent cyclones had hit hardest and washed away homesteads, with maps and stories provided by the media, all indicating that the centers of these storms had luckily only scratched along Sagar's edges instead of hitting with full force.

If protection here came through the direct interventions of the divine yogi, it also operated indirectly, I suggest. After long elaborations on the genesis of cyclonic pressure fields, many of my interlocutors would, for instance, point to how deep their lands lay. Most storms, I learned in conversations across the length and breadth of the island, would simply pass Sagar, flying high in the sky, too high to even touch this low-lying landmass. "This is the deepest place of the earth," said Anil Patro once, mirroring the voices of many others. "The storms spare Sagar because Sagar is situated so deep that the storms rush high above it without touching it." It took me a while to understand what was at stake here. On one level, such statements evidence the considerable "disaster gap" (Pfister 2009) Sagar experienced before Aila in 2009, and between Aila and Fani in 2019 and Amphan in 2020. Islanders were aware of horrific cyclones regularly hitting East India and Bangladesh in recent history, they just believed Sagar to be exempt from a similar fate. Referring to storms hovering high above the island, without ever touching it, draws on the idea that the place where Kapil Muni resides is literally the lowest place on earth. Remember, it was excavated by King Sagar's 60,000 sons, who raged wildly across the continent, tearing open the netherworld in an aggrieved hunt for their father's stately horse (see chapter 4). Situated deep in the netherworld—a cave at the bottom of the earth's deepest pits—the island is out of reach of storms rampaging across the world's surface.

Less often, islanders would rely on letter-based models to account for the curious quiet the island enjoyed in an area otherwise churned by cyclones. Ganesh, a farmer and father of four, who had been taken to Colony Para by his parents, mobilized an account that combined meteorology with scripture. Cyclones moved through the bay along trajectories, he said, that resembled the character *ba* (ব) of the Bengali script. Some, like the 1999 supercyclone, take the lower route, hitting Orissa, while others take the straight route and end up hitting land in Bangladesh. In most cases, it is one of these two, and Sagar sits, comfortably, in the blank space between them.

Ganesh's theory involved alliteration, using script as a tool for rendering statistical variation understandable. References to deep lands and storms hovering high above it, however, situate islanders in a scripture-based cosmology that provides protection in itself. In this view, there is no need for the divine yogi to swing into action. In other words, dwelling alongside Kapil Muni in his cave holds promise on its own: it means being beyond the reach of many storms.

But while the island's position in the netherworld was seen to account for the absence of storms, it also entailed danger. Here I am not referring to uncanny dimensions of living in the netherworld, since risks involved in cohabiting spaces with subterranean monsters, of which the scriptures speak at length (Valmiki 2005), never cropped up. I rather refer to how this revered depth puts the fragile island into the reach of river and sea, and enhances its vulnerability to floods and erosion. This materializes when islanders speak of the island as a "flood zone" (*bonya elaka*). Safety from one hazard is caveated by vulnerability to another.

Some suggested Kapil Muni could tamper with high tides as well. Devnona Jana, for instance, explained that the divine yogi "never let the spring tide rise higher than to the knee." And many more insisted that the island as a whole would continue withstanding the waters with the help of divine protection. After all, this was not just any place, as I was told over and over again, but Kapil Muni Thana, his worldly realm and territory (Jacobsen 2012, 106–107)—literally the territory of Kapil Muni's police station (*thana*). Since Sagar was "the place where Kapil Muni resides," old Sheikh Shor told me, "it will not sink." Reading Sagar as the territory and jurisdiction of Kapil Muni is not a figure of speech. It connects the present to Hindu cosmology, and is bolstered by legal regulations that ban the sale of liquor within a twelve-kilometer radius of the temple. Importantly, it also underpins more optimistic takes on futures in a turbulent estuary. Across all my field sites, a majority of islanders felt that even in the face of current onslaughts, the overall existence of the island was not at stake. Tides would break through embankments and gobble up homes, and storms would crash down trees or disrupt services, but the island still had a future.

The benefits of dwelling with Kapil Muni in his homeland were manifold. Even while the divine yogi rarely ever figures in domestic or public ritual among islanders (Jacobsen 2008), his guiding presence, so to say, is of critical value for resuscitating hope while inhabiting the landmass and looking out toward the future.

Such imagery resonates with mythical accounts of the protection from encroaching waters by hypermasculine actors. Hindu scriptures proclaim that when Ganga finally reached the seashore and began emptying herself into the ocean, the sea level rose and lands in today's Kerala were swamped (Eck 1998, 72–73). Responding to the plights of settlers in this now sunken land, the mighty Brahmin Parashurama stepped in, himself an avatar of Vishnu, just like Kapil Muni. Pointing his bow at the sea god Varuna, a virile Parashurama threatened and ordered him to withdraw the excess water from the land. Terrorized, Varuna paid heed and freed the land by draining the waters. Famous across India, the story is one of very few hints at the dangers of the celestial river—even if the danger here appears to be indirect, driving sea level rise on a distant shore.

It is tempting to read this as a premodern forecasting of the Anthropocene, where human activity—this time mediated by divine forces acting at the behest of humans—irrevocably alters the environment and occurs as a geological force. More importantly for my purposes here, however, these stories articulate a theme—virile, masculine deities keeping female deities in check for the sake of humanity.

Similar takes on masculine protection from excess waters also pervade the shores of the Western Bengal Delta. The Muslim saint Pirbaba, for instance, understood to reside in Hijli Sharif, on the other side of the Hugli, is revered as a warrior-like figure. Stories of how he drove the British out from his land, i.e., from his tomb complex, by materializing tigers are hugely famous locally. On Sagar, bazaar prints referring to this incident adorn the homes of Muslims and Hindus alike. Islanders also admire Pirbaba for protecting his followers and his realm from watery threats. He is seen to rescue fishermen lost at sea from drowning and to push back floods from entering his dominion (also see the hagiography in Giri 2007, 35). Once again, we see a hypermasculine divine figure subduing a hostile sea.

In Bangladesh's deltaic tracts, not very far from Sagar, people address Khwaj Kizir, or Jol Devota, as a way of seeking influence on the unruly waterscape. Khwaj Kizir is a Muslim saint, venerated as the king of waters (Zaman 1999; Khan 2016). Jol Devota, on the other hand, literally translates as "water deity" in the Hindu pantheon. In order to avert erosion, folks turn to either of them in prayer (Zaman 1999, 200). In a reversal of such a stance, islanders on Sagar instead seem to resort to lords of the land, such as Kapil Muni and Pirbaba, pleading for safety from the maddened female divinity that is the river surrounding their home islands.

All these iterations of hypermasculine protection come with their limits. At their best, they are concerned with pushing waters back from one's dominion. At their most basic, they aim to temper devastations and secure the promise of overall protection of the island's landmass, rather than a complete end to *bhangon*. It is hard to deny that the divine protector appears to be struggling with the power of the sea. After all, Gangasagar's main temple housing Kapil Muni is not the first one of its kind. It was built after its predecessor had succumbed to the encroaching sea. Conversations about what might be called the displacement of one deity by another led to uncertainty and silence, presumably because they indicated the limits of the landed god's power and thus the limits of his promises. Priests running Gangasagar's main temple remained tight-lipped and shunned my inquiries into the displacement of the earlier temple. Among my Bengali interlocutors, an uncomfortable silence set in whenever we hit the subject. On the one hand, the micromigration and resettlement of divinity come as little surprise in a world animated by erosion and transformation. On the other, however, a retreating temple not only emblematically intensifies the predicament many islanders find themselves in, it also clouds divine agency in uncertainty.

As with Ganga, it proved difficult to discuss these matters with my interlocutors. Silences ensued, ambivalences took hold. Such discussions could slip into blasphemy. It seemed risky to question the agency and power of the very divinity one depended upon. It might also prove helpful to invest into hopes against all odds, mobilizing divinity alongside political connection in order to weather an uncertain present.

For people residing at these eroding edges, notions of the island's ultimate protection amounted to little immediate relief. Such assessments never stood in the way of demanding massive investments in coastal protection or positioning battered islands as crumple zones, suffering for the sake of others (Harms forthcoming). Even in and around Colony Para, where *bhangon* seemed to be arrested for the moment, and where the provisions by the divine yogi were relied upon to ensure a stable future, darker shadows and troublesome assessments were lurking.

Across the years of my research, visions of masculine saviors have been complemented by the motherly figure of Didi—Mamata Banerjee, who has dominated electoral politics in West Bengal since 2011. I have noted above how individual embankments are rendered in the language of gifts bestowed by Didi, and shown how the development of Gangasagar appears to have been bolstered when she took over the state administration (chapter 6).

Whether or not Mamata's promises were hollow, as critics argue, despite mounting unease with her party's rule, her reign saw development of the pilgrimage center of Gangasagar. Combining drives to bolster Hindu heritage with attempts to establish seaside resort tourism made the sacral center a focal point of rural development. Arguably, the whole island benefitted from this revived interest. Infrastructural development, such as wide roads and ferry connections, are clearly designed for pilgrimage traffic but smoothen journeys beyond the days of the festival. Similarly, electricity cables now reach across the isthmus between Sagar and the mainland, providing round-the-clock electricity in the pilgrimage center and along the artery road leading toward the main temple. What is more, the pilgrimage center itself has seen massive investment lately in the form of concrete structures for markets selling religious paraphernalia, which also double up as shelter from floods and cyclones.

The way Mamata distributes her gifts produces winners and losers, mirroring how development in general diffracts geographies and inscribes unevenness and (un)care. It therefore is not simply a case of the state doing its duty, as some of my interlocutors would have it. Didi's massive infrastructure projects on these shores—and not on others—instantiate a break with what her party decried as disregard of Hindu tradition and heritage by her forerunners and chief adversaries from the Left Front. Here we see her government's alignment with North Indian mainstream visions of Hinduism. Positioning the pilgrimage festival as a segment of so-called national heritage and a cornerstone of essentialized Hindu identity is deeply problematic—not the least for sidelining ambivalent takes on fluid divinity. Yet it also, I would suggest, buttresses a sense of ultimate protection among islanders. In bringing up infrastructure, Mamata recognizes the special status of the island, in itself a feat, and adds state protection on top of it.

State investments thus further inscribe the sense that Sagar inhabitants are protected denizens of Kapil Muni's cave. Politics and ritual entangle in ways that provide for a future on shaky grounds. In the figure of Didi, protection by hypermasculine actors is complemented by the protection of an assertive mother. Pointedly, then, it can be argued that when waves or engineers do what gods do, the opposite also holds true: politicians seem to realize the safety provided by the divine hermit.

Pointing Fingers in the Anthropocene

This chapter has pieced together the juggling act faced by islanders trying to make sense of the eroding shores of the western Sundarbans. It has shed light on how distributed disasters are perceived, and how a sense of ultimate protection is salvaged, even in the aftermath of displacement. I make the case for one crucial contrast. While disasters the world over tend to be narrated in somewhat unified terms, among islanders such narrations are marked by ambiguities and profound uncertainties. They diverge among people and groups, swaying between the religious ideas and practices underpinning society and the desire for better protective infrastructure against ruthless waves. This is not to say that all my interlocutors voiced such diversity, but that shifting rhetoric and entangled imageries abound. What I have accounted for is the entanglement of discourse and imagery informing the majority of their narrations. These entanglements are closely related to a set of uncertainties mapped across the notorious nature/culture divide, specifically about the exact causes of coastal erosion. Given the presence of divinity worked through by human interventions near and far, I have argued that these uncertainties are, in their respective ways, related to both the texture of the disaster and the materiality of brackish environments.

The fact that the disaster is a distributed one is crucial. It postpones, dispenses, and defers attributing responsibility. The texture of the waterscape itself adds to the conundrum. Islanders are acutely aware of its multiple constituents and the diversity of forces that impact the shape of every wave and current. Clearly visible entities, such as cargo ships, meet invisible interventions, such as excess emissions driving anthropogenic sea level rise. Cosmological orders intersect, when islanders relate to the water simultaneously as a body of divinity and as a port channel. They intersect when evaluating the impact of surges in terms of meteorological fields or moon-driven tides, and in terms of the agency of divine sages. Similarly, the slow shifts in the estuary, be they the grace of divinity or of the Anthropocene, meet rather rapid-onset shifts, following, for instance, the development of coastal protection measures.

It is perhaps little wonder that Juddhistir Jana was left a little stumped when asked about the behavior of the waters around Sagar Island. Here, hydraulics, politics, and divinity are deeply intertwined, and islanders are left trying to make sense of the conundrum. In tracing how waves and currents are evaluated on these disappearing lands, I have established two related claims. I first demonstrated how religion and politics converge in narratives of blame that do consist of moralized accounts of wrongdoing. Gods, engineers, and politicians converge as rather self-absorbed, inaccessible, and frequently careless actors. Second, I teased out coexisting imageries of protection that are decidedly gendered and signal ultimate survival amid otherwise bleak futures. Debates about what permeates water, and what water is, turned into opportunities for critiques of politicians and gods. The modality of this critique, however, is telling. While critiquing politicians has a long history and is, at times, a pastime of sorts in rural Bengal—if not a passionate and outspoken activity (Ruud 2000; Spencer 2007)—to criticize gods remains risky, and any blame is consequentially voiced hesitantly.

To shift back and forth between hydraulics, politics, and divinity leaves us firmly in a terrain of ambivalence. Islanders and researchers alike are stuck within a constrained discursive field that appears to be the only viable and safe one up against an overtly powerful river and a hegemonic concept of the riverine goddess who underpins everyday existence in the shadow of a pilgrimage center. Any significant capacity or willingness to help is again invested largely in virile actors, be it a divine yogi or a chief minister known for her prowess and cunning habits. Local political ecologies—that is, islanders' assessments of ecology and power—thus oscillate between modestly heretical takes and more outspoken blame of secular politicians.

Life on the banks of Ganga and the delta beyond has lots to offer inquiries at the intersection of religion and disaster, and the study of sacred landscapes. It shows that environmental hazards may impact ritual tenets and practices, all the way to reworking the character of divinity itself. In this sense, the study of environmental displacement has something to say about how people deal with degrading environments, as well as how disruption and despair impact longstanding beliefs. It adds layers, I suggest, as ambiguity is expressed in the face of powerful mainstream visions, stopping short of outright blasphemy. Approached differently, such ambiguity heightens the difficulties displaced islanders face, for they have to negotiate not only treacherous waters but also the treacherous interpretations of what has befallen them.

Chapter 8

Conclusion

Mosaics and Futures

One day, I happen to be caught in the middle of a conversation between a mid-aged couple and a lady, and I cannot but follow. We're on board the privately operated wooden trawler that goes back and forth between Sagar and Ghoramara three to four times per day, tides and daylight permitting. People gather in front of the boat, squatting on planks under their umbrellas, holding fast to groceries, patient records, or suitcases. Even while the diesel engine is working at maximum, conversations still unfold at the front. The couple is sitting next to me on my left, the lady squats on the planks to my right. I don't know either of them, but they clearly have known each other for a long time, if obviously not very well, and they engage in a catch-up of sorts.

As the trawler pushes across the river, the couple elaborates on where they are now. Ferry crossings are exercises in reflection, and theirs is a story of loss, longing, and nostalgia. They have bought a small piece of land on the mainland nearby, some forty kilometers away, a new home to which they are now returning after a short visit to the island. In a way, they have made it. They succeeded, one could say, in moving out. They secured a plot that appears to be beyond the reach of *bhangon*. But on another level, they seem to have lost everything. "There is nothing," the man explains. No field to plough and no pond to have fish from. "How much money is wasted," the women chips in, "on buying rice on the market!" Safety is gained, and rural well-being lost forever, or so it seems.

Accounts of rising waters and land subsidence now customarily end with invocations of doom. Social scientists frequently refer to the unfolding sacrifice of land and people, or to the specter of dying cities and soon to abandon coasts (see, for instance, Pilkey, Pilkey-Jarvis, and Pilkey 2016). Journalists find futures impossible on sinking islands, warning of massive exodus (see, for instance, Narayanan 2015). Politicians from small island states never tire of warning about the eradication of their nations from the face of the earth (see, for instance, Campbell and Bedford 2022). All of this has good reason. Outlooks are bleak, and the ranks of people displaced by anthropogenic shifts and ruptures along the coast are set to rise in the coming years and decades.

Nothing of this is entirely new. And not all is lost, nor will it ever be. The conversation on the ferry highlights provisional engagements with coastal erosion, fueled by the determination to stay put and to survive and strive against all odds. To frame the couple returning to their islet as "too poor to move" or as members of "trapped populations" (L. M. Hunter 2005) is only part of the story. They didn't have the means to seek their fortunes across borders, nor the inclination to move into terrain far away from this sea to be fully out of the reach of brackish waters. It is questionable whether they would ever have the inclination to do so. In content and context, the conversation also highlights some of the intricacies of securing futures among shifting seas.

It wouldn't do justice to note that the couple was simply returning to one place with its particular troubles from another place and its troubles. To be sure, their new home was blighted by population density and by its absence of space to engage in subsistence farming, in addition to soaring prices at the urban fringe. Yet, the new place seemed to provide a hold on permanence that the shrinking islet had ceased to offer long ago. I have argued in this book that the experience of coastal erosion makes deliberations on whether this is an involuntary or merely an induced displacement ultimately redundant. Land disappearing, and thereby literally removing the ground from under one's feet, makes mobilities unavoidable and displacements an imminent experience. It follows that coastal erosion does not simply sit as one factor among many others, all threatening the survival and well-being of marginalized populations, but that coastal erosion is undoing land-based forms of life by removing their underpinning.

The couple's sojourns represent types of involuntary mobility that lead away from eroding islands yet remain within high-risk areas within the coastal zone. Their mobilities certainly transgress what I have called micromigrations by leading into fairly distant terrain beyond the reach of tight networks of kin, friends, and neighbors and featuring decidedly unfamiliar traits. I am left wondering if to the couple the hazards of moving out were balanced by the promise of permanence the new place held. It promised to remain, to stay above waters, and to be untouched by coastal churning for the foreseeable future.

There is no way of knowing the future. Still, I think that the future vision implicit in the couple's move appears much more realistic than the visions implicit in those maps illustrating sea level rise as sweeping incursions of the Bay of Bengal into the delta. As noted, such scenarios fail to take adaptive interventions into consideration that may contain water or push land above the tide line. In other words, coastal erosion will in the future proceed with greater velocity and with greater social costs in areas exempted from the care provided by the state or corporations. Urban conglomerates stand better chances of persisting, with rural terrain receiving much less of the investments and the clout necessary to withstand marine incursions.

Replacing the scenario of delta drowned wholesale with that of fortified urban islands amid abandoned rural coasts is again oversimplifying. In his research on urban flooding, Lukas Ley shows that marginalized settlements may not only house terrain most vulnerable to saltwater incursions, but their populations are most likely to be left alone in them. State uncare feeds into self-styled edgework and exhaustion in the sinking poor neighborhoods of flourishing coastal cities (Ley 2021). There are other pieces to this mosaic as well. On rural, altogether underprotected shores, individual coastal stretches receive substantial care in the form of concrete dikes. It remains up for debate if and for how long such dikes can contain the sea, and whether such noncomprehensive approaches actually are helpful at all. At the very least, they flag areas of state concern and articulate abandonment everywhere else.

Registering the mosaics of coastal protection is one step of gauging futures. Another is placing islands within histories, scripture-based or otherwise. Here I am referring not to notions of antiquity as a way of emphasizing endurance and suggesting continued existence. I rather gesture at the role of specific islands within scripture-based notions of protection by deities and saintly figures. Ultimate protection and state care may align very well in practice, especially in a climate of developing an island harboring a pilgrimage center by governments subscribing, as noted, to versions of Hindu nationalism and sacral geographies sustained by pilgrimage sites.

Betting on the village my co-travelers on the ferry to Ghoramara now call home, and investing in a future there, may also have something to do with the size of the landmass it sits on. After all, it is not only the personal or otherwise gauged experience of erosions that renders islands vulnerable, but also their very form. Of low elevation and small perimeter, Ghoramara appears barely afloat on the coastal seams of a vast estuary. Its fragility is visible to the eye at most times.

Sagar, on whose northern tip the couple boarded the ferry, also shone by its sheer size. I did not bring this up with the couple, but, to many of my interlocutors, this comparatively vast island seemed "too big to fail." In adapting this phrase from debates around the necessity for the state to buy out banks in the wake of the 2008 financial crises (Riles 2013), I am not suggesting that state actors eventually will step in and provide protection, nor that islanders bet on this kind of protection. I use it rather to capture how the size of a specific landmass might in itself seduce people to believe in its endurance. Hard to pin down, such beliefs collate with evidence of state investing, at least in pockets, and with deep-seated notions of ultimate protection by a powerful yogi, evident in the social worlds within and around Colony Para. To consider oneself safeguarded by divine agency and to note beneficial development initiatives on a vast landmass certainly underpins optimistic takes of the future. It helps to keep the bleakest future visions at bay. Envisioning futures buttressed by such beliefs has, of course, real-life effects. They animate people securing futures by building homes, struggling for embankment protection, or lending their own bodies as additional layers to failing dikes when yet another cyclone makes landfall. In their minute and often barely visible ways, these efforts intervene into the waterscape and impact the contours of islands. They skew the mosaic, impacting hydrological relations by holding firm to the land through a combination of minor infrastructures, future reckonings, and solidarities built on shared histories.

These constellations of loss and hope, the material realities they intervene in, as well as the relations they sustain, are of import to environmental theory in this moment of planetary crises. Across the chapters of this book, I have demonstrated how the experience of coastal erosion enforces rethinking the conceptual apparatus of disaster studies and of humanitarian interventions by acknowledging what I have introduced as distributed disasters. This is not a purely academic matter, but a first step toward devising appropriate responses.

I have shown how sea encroachment meets marginalization on crowded shores. Finding ways out of the gridlock is an enormous task. Marine scientists have warned for centuries that the sea-facing sections of the archipelago south of Kolkata is unfit for human habitation (see, for instance, Fergusson 1863)—it is too low and vulnerable to settle comfortably, while settlement disrupts the rising of the land by way of accretion, keeping it perpetually at risk. In private communication, experts working at the overlap of academic research, humanitarian response, and activism told me that there is ultimately no way around resettling the whole delta-front population elsewhere. Obviously, uprooting and resettling a population in the millions is not a realistic option. Even if there were the political will to find them a place and to secure their dignity, such a move would involve a great deal of violence. Formulating such "solutions" also denies the mosaic of coastal protection I have engaged in this book, and the adaptive interventions into the waterscape pursued at different levels. Considering the broader histories of involuntary migrations that inform experiences during present-day environmental displacements adds a crucial layer for understanding decision-making at the crumpling, densely populated, and land-scarce shores. Acknowledging these dimensions, and placing them within uneven capabilities to rework and fortify the coastal seams, opens up vistas on coastal resilience and ways to support it.

Introducing the notion of distributed disasters has enabled me to explore, and situate, specific forms of environmental suffering at Bengal's eroding coastal seams. The disappearance of land, and the suffering it entails, provide for particularly stark forms of environmental degradation. Coined to make sense of these, the notion of distributed disasters promises to be illuminating elsewhere too. It enables capturing the experience of living in environments whose habitability becomes inverted by a combination of gradual shifts and rhythmic disruptions. Foregrounding, as I have done here, the distributed nature of these processes enables understanding the way such processes impact mundane practices and require specific modes of endurance. It also buttresses attempts to reach toward environmental justice by developing response mechanisms taking into account, among others, the small-scale yet widely scattered, rhythmic, and existentially threatening quality of environmental degradation.

Detailing how islanders experience the suffering induced by coastal erosion against the background of marginalization and abandonment is critical for coming to terms with some of the troubling outcomes of the Anthropocene. Turning to endurance and exhaustion, to displacement and resettlement, helps us to understand how people sustain environmental degradations that are at once barely visible and existential along some of Asia's shorelines.

Important as it is to provide ethnographic depth on the situated experience of coastal erosion on three remote islands, these explorations have salience beyond that. I am using the label of the Anthropocene here deliberately. As a geological epoch, it implies long temporal arcs and the entanglement of political economy and ecology over time. It also highlights extended moments of experimentation, of a prolonged sense of urgency to come to terms with shifts that are simultaneously profound and hard to pin down. Looking back in time and detailing how some of the projected shifts have been with us and been navigated for quite a while now is an essential step of settling into the Anthropocene.

In the resettlement operations that I have recounted here, many of the concerns voiced by scholars working on displacement and rehabilitation seem to fade into the background, such as transnational migration, ethnic belonging, or camp life. My account foregrounds how, on the one hand, concrete resettlement drives are driven by, and insert into, rather constrained territorialized networks. On the other hand, I have shown how such networks and, more broadly, how imaginaries of land-based, firmly anchored visions of the good life fare in zones of rampant coastal erosions. Exploring life along flailing embankments and in resettlement colonies highlights well-developed concerns of political ecology and critical development studies—such as questions of access to land or water and to welfare benefits—and reinserts them into debates on the coastal future on crowded, neglected shores.

Considering the cyclicities and circuits I have detailed here, neither erosion nor relocations are entirely novel and always appear as actualizations of preceding developments. Yet, the circuits appear to be tightening in some respects, where expansion and settlement of swamps has long come to an end; and they do increasingly also overlap with circulatory forms of labor migrations or with the shifting cycles of rapid-onset disasters, such as surges, or slow-onset disasters, such as droughts.

These cyclicities also get entangled with the rhythms and patterns making up the waterscape that is the mouths of the Ganga. If the fluid body of divinity gets worked up by rotors, and if some of the currents that embody her playfulness are intensified by subsurface walls, what is this thing called Ganga then? In this book, I have stayed truthful to hesitant and ambiguous takes on the nature of water, and invested in drawing attention to specific forms of relations entertained through muddy, brackish waters in this domain of *bhangon*. I have uncovered a sense of not being in control, of overwhelm. This seems to be characteristic for water relations sustained through estuaries or seas, animated, as they are, by diverse peoples, material cycles, or scaled interventions. Coming to terms with vast waters and their heterogeneous, ultimately obscure drivers often means, as I have shown, highlighting near or mid-range interventions and pinning both blame and hope onto these. Conversations were ripe with allusions to subsurface walls, divine protectors, and container ships. But while all these were always in hindsight, and in many ways easier to engage with than, say, climate change, in the eyes of my interlocutors, they remained beyond control. Indeed, the interventions appear to be rather easily adjustable. Technically, subsurface walls could be dismantled and marine traffic could be diverted. Yet, within the political conditions of the present, which prioritize trade and industrial development over agrarian lives and marginal islands, such adjustments are unlikely at best.

Joining islanders debating the repercussions of redirected flows or the nature of divine waters, as well as considering how coastal erosion inserts into marginalization, has been an exercise in situating, and provincializing, climate change. This is not to deny that global warming takes its toll on these shores, and has been for quite a long while already, nor that we should learn from experiences here what coastal futures will demand as temperatures rise globally. Instead, it is meant to draw attention to the ways global warming inserts into already fraught and oftentimes volatile conditions. Rising seas add pressure to redirected flows, and the intensification of coastal erosion adds pressure in impoverished homes. As a concept, climate change enters an already multilayered set of abstractions that people rely on to explain weather patterns, environmental degradations, and potential futures.

Seen from Sagar's embankments climate change sits among a number of interventions into the waterscape that greatly impact islanders' well-being, yet onto which they have little to no control. This reorients the debate on climate change and responsibility. Indian environmentalists have long made the case for the unequal responsibilities of causing climate change in a postcolonial world (Agarwal and Narain 1991). Attending to waves and currents, and winds and rains, places these inequalities alongside unequal provisioning and the violence built into state-sanctioned development trajectories.

It is ironic that the narratives of Ghoramara's notoriety as a hotspot of climate change endangerment frequently gloss over these inequalities. Using the fate of the island as a lens to illuminate the unequal burden of climate change is important. But it should not divert attention from interventions within the immediate waterscape—from docks to thorough embanking across the estuary all the way to the Farakka Barrage upstream—that spell out barely visible, slow forms of violence as well. In this view, provincializing works toward situating responsibilities.

Accounting for the dynamic and mosaiclike quality of crumpling shores requires us to take human interventions seriously. These might be miniscule, such as the everyday practice of plastering the walls of mud houses that the state abandoned to the floods but which still serve as homes. They might be ephemeral, like the passing of ships whose rotors arguably churn the waters. And they might be too big to perceive, like the shifts in climatic conditions, or illegible to secular debates, such as attempts to pacify the other Ganga. Not all of these are interventions into the material fabric of the delta, and even less so into its soils and protection infrastructures. Some of them rather impact the ways islanders place themselves on the coastal seams. This has consequences, I have argued across this book, for situated practices of envisioning and implementing futures, which in turn animates interventions into the material ecology. Situating islanders' life histories may in this view enable our understanding of modes of endurance and hope along dramatically eroding shores and allow us to move beyond discourses of victimization and haplessness. It is at the intersections of overwhelm and empowerment that coastal futures emerge—on the Bengal delta front and beyond. Considering how politics, matter, and visions all intermingle in muddy formations will contribute toward achieving just transformations.

References

Adam, Hans Nicolai. 2015. "Mainstreaming Adaptation in India—the Mahatma Gandhi National Rural Employment Guarantee Act and Climate Change." *Climate and Development* 7 (2): 142–152.

AFP. 2019. "Indian Island Residents Vote with Sinking Hearts amid Climate Change Fears." *Straits Times*, May 19, 2019. https://www.straitstimes.com /asia/south-asia/indian-island-residents-vote-with-sinking-hearts-amid -climate-change-fears.

Agarwal, Anil, and Sunita Narain. 1991. *Global Warming in an Unequal World.* New Delhi: Centre for Science and Environment.

Agrawal, Arun, and K. Sivaramakrishnan. 2000. "Introduction: Agrarian Environments." In *Agrarian Environments: Resources, Representation, and Rule in India,* edited by Arun Agrawal and K. Sivaramakrishnan, 1–22. Durham, NC: Duke University Press.

Ahmed, Sara, Anne-Marie Fortier, and Mimi Sheller, eds. 2003. *Uprootings/ Regroundings: Questions of Home and Migration.* Oxford: Berg Publishers.

Ajibade, Idowu Jola, and A. R. Siders. 2021. *Global Views on Climate Relocation and Social Justice: Navigating Retreat.* London: Routledge.

Ajibade, Idowu, Meghan Sullivan, and Melissa Haeffner. 2020. "Why Climate Migration Is Not Managed Retreat: Six Justifications." *Global Environmental Change* 65:102187.

Albinia, Alice. 2009. *Empires of the Indus: The Story of a River.* London: John Murray.

Albrecht, Glenn A. 2019. *Earth Emotions: New Words for a New World.* Ithaca, NY: Cornell University Press.

Alley, Kelly D. 1994. "Ganga and Gandagi: Interpretations of Pollution and Waste in Benares." *Ethnology* 33 (2): 127–145.

———. 2003. *On the Banks of the Gaṅgā: When Wastewater Meets a Sacred River.* Ann Arbor: University of Michigan Press.

Anand, Nikhil. 2017. *Hydraulic City: Water and the Infrastructures of Citizenship in Mumbai*. Durham, NC: Duke University Press.

Appadurai, Arjun. 1988. "Putting Hierarchy in Its Place." *Cultural Anthropology* 3 (1): 36–49.

Arnold, David. 2006. *The Tropics and the Traveling Gaze: India, Landscape, and Science, 1800–1856*. Seattle: University of Washington Press.

Asad, Talal. 1993. *Genealogies of Religion: Discipline and Reasons of Power in Christianity and Islam*. Baltimore, MD: Johns Hopkins University Press.

———. 2007. *On Suicide Bombing*. New York: Columbia University Press.

Ascoli, Frank David. 1921. *A Revenue History of the Sundarbans from 1870 to 1920*. Calcutta: Bengal Secretariat Book Department.

Augé, Marc. 2004. *Oblivion*. Minneapolis: University of Minnesota Press.

Auyero, Javier. 2012. *Patients of the State: The Politics of Waiting in Argentina*. Durham, NC: Duke University Press.

Auyero, Javier, and Debora Alejandra Swistun. 2009. *Flammable: Environmental Suffering in an Argentine Shantytown*. New York: Oxford University Press.

Babb, Lawrence A. 1975. *The Divine Hierarchy: Popular Hinduism in Central India*. New York: Columbia University Press.

Badiani, Reena, and Ablir Safir. 2009. "Circular Migration and Labour Supply: Responses to Climatic Shocks." In *Circular Migration and Multilocational Livelihood Strategies in Rural India*, edited by Priya Deshingkar and John Farrington, 37–57. New Delhi: Oxford University Press.

Baer, Hans, and Merrill Singer. 2014. *The Anthropology of Climate Change: An Integrated Critical Perspective*. London: Routledge.

Baldwin, Andrew, and Bruce Erickson. 2020. "Introduction: Whiteness, Coloniality, and the Anthropocene." *Environment and Planning D: Society and Space* 38 (1): 3–11. https://doi.org/10.1177/0263775820904485.

Ballestero, Andrea. 2019. *A Future History of Water*. Durham, NC: Duke University Press.

Bandyopadhyay, Sunando. 1997. "Natural Environmental Hazards and Their Management: A Case Study of Sagar Island, India." *Singapore Journal of Tropical Geography* 18 (1): 20–45.

———. 2008. "Nayachar: Factories in Place of Mangroves." In *Nandigram and Beyond*, edited by Gautam Ray, 199–224. Kolkata: Gangchil.

Bandyopadhyay, Sunando, Nabendu Sekhar Kar, Sayantan Das, and Jayanta Sen. 2014. "River Systems and Water Resources of West Bengal: A Review." *Geological Society of India Special Publication* 3 (2014): 63–84.

Banerjee, Abhijit Vinayak, Pranab Bardhan, Kaushik Basu, Mrinal Datta Chaudhury, Maitreesh Ghatak, Ashok Sanjay Guha, Mukul Majumdar, Dilip

Mookherjee, and Debraj Ray. 2007. "Beyond Nandigram: Industrialisation in West Bengal." *Economic and Political Weekly* 42 (17): 1487–1489.

Banerjee, Anuradha. 1998. *Environment, Population, and Human Settlements of Sundarban Delta.* New Delhi: Concept.

Bankoff, Greg. 2003. *Cultures of Disaster: Society and Natural Hazard in the Philippines.* London: RoutledgeCurzon.

Basso, Keith H. 1996. *Wisdom Sits in Places: Landscape and Language among the Western Apache.* Albuquerque: University of New Mexico Press.

Basu, Jayanta. 2020. "Cyclone Amphan: How Brick Homes, Aila Embankments Saved the Day at Sundarbans Village." *Down to Earth*, June 9, 2020. https://www.downtoearth.org.in/news/natural-disasters/cyclone-amphan-how-brick-homes-aila-embankments-saved-the-day-at-sundarbans-village-71655.

Basu, Subho, and Auritro Majumder. 2013. "Dilemmas of Parliamentary Communism." *Critical Asian Studies* 45 (2): 167–200. https://doi.org/10.1080/14672715.2013.792569.

Bauer, Andrew M., and Mona Bhan. 2018. *Climate without Nature: A Critical Anthropology of the Anthropocene.* New York: Cambridge University Press.

Baviskar, Amita. 2003. "Introduction: Confluences and Crossings." In *Waterlines: The Penguin Book of River Writings*, edited by Amita Baviskar, ix–xiv. New Delhi: Penguin India.

———. 2004. *In the Belly of the River: Tribal Conflicts over Development in the Narmada Valley.* New Delhi: Oxford University Press.

Bear, Laura. 2007. "Ruins and Ghosts: The Domestic Uncanny and the Materialization of Anglo-Indian Genealogies in Kharagpur." In *Ghosts of Memory: Essays on Remembrance and Relatedness*, edited by Janet Carsten, 36–57. Malden, MA: Blackwell Publishers.

———. 2012. "Sympathy and Its Material Boundaries: Necropolitics, Labour and Waste on the Hooghly River." In *Recycling Economies*, edited by Catherine Alexander and Josh Reno, 185–203. London: Zed Press.

Bellacasa, Maria Puig de la. 2017. *Matters of Care: Speculative Ethics in More than Human Worlds.* Minneapolis: University of Minnesota Press.

Berliner, David. 2005. "The Abuses of Memory: Reflections on the Memory Boom in Anthropology." *Anthropological Quarterly* 78 (1): 197–211.

Bhaṭṭācārya, Tarunadeba. 1976. *Gangasagar Mela: A Pilgrim's Guide.* Calcutta: Office of the District Magistrate.

Bhattacharyya, Debjani, and Megnaa Mehtta. 2020. "More than Rising Water: Living Tenuously in the Sundarbans." *The Diplomat*, 2020. https://thediplomat.com/2020/08/more-than-rising-water-living-tenuously-in-the-sundarbans/.

Bille, Mikkel, Frida Hastrup, and Tim Flohr Sorensen, eds. 2010. *An Anthropology of Absence: Materializations of Transcendence and Loss.* New York: Springer.

Birla, Ritu. 2009. *Stages of Capital: Law, Culture, and Market Governance in Late Colonial India.* Durham, NC: Duke University Press.

Black, Richard. 2001. *Environmental Refugees: Myth or Reality?* Geneva: United Nations High Commissioner for Refugees.

Black, Richard, W. Neil Adger, Nigel W. Arnell, Stefan Dercon, Andrew Geddes, and David Thomas. 2011. "The Effect of Environmental Change on Human Migration." *Global Environmental Change* 21 (Supplement 1): 3–11.

Black, Richard, Stephen R. G. Bennett, Sandy M. Thomas, and John R. Beddington. 2011. "Climate Change: Migration as Adaptation." *Nature* 478 (7370): 447–449.

Black, Richard, and Michael Collyer. 2014. "'Trapped' Populations: Limits on Mobility in Times of Crisis." In *Humanitarian Crises and Migration: Causes, Consequences and Responses,* edited by Susan F. Martin, Sanjula Weerasinghe, and Abbie Taylor, 287–305. London: Routledge.

Blaikie, Piers, and Harold Brookfield. 1987. *Land Degradation and Society.* London: Routledge.

Blair, Harry W. 1990. "Local Government and Rural Development in the Bengal Sundarbans: An Inquiry in Managing Common Property Resources." *Agriculture and Human Values* 7 (2): 40–51. https://doi.org/10.1007/BF01530435.

Blanchet, Thérèse. 1984. *Meanings and Rituals of Birth in Rural Bangladesh: Women, Pollution, and Marginality.* Dhaka: Dhaka University Press.

Blumenberg, Hans. 1997. *Schiffbruch mit Zuschauer: Paradigma einer Daseinsmetapher* [Shipwreck with spectator: Paradigm of a metaphor for existence]. Frankfurt: Suhrkamp.

Bose, Debaki, dir. 1959. *Sagar Sangame* [The holy island]. Feature film.

Bose, Sugata. 1993. *Peasant Labour and Colonial Capital: Rural Bengal Since 1770.* Cambridge: Cambridge University Press.

———. 2007. *Agrarian Bengal: Economy, Social Structure and Politics, 1919–1947.* Cambridge: Cambridge University Press.

Brammer, Hugh. 2014. "Bangladesh's Dynamic Coastal Regions and Sea-Level Rise." *Climate Risk Management* 1 (Supplement C): 51–62. https://doi.org/10.1016/j.crm.2013.10.001.

Bresnihan, Patrick, and Arielle Hesse. 2021. "Political Ecologies of Infrastructural and Intestinal Decay." *Environment and Planning E: Nature and Space* 4 (3): 778–798.

Brockington, Dan. 2002. *Fortress Conservation: The Preservation of the Mkomazi Game Reserve, Tanzania*. Oxford: James Currey Publishers.

Bryant, Raymond L., and Sinéad Bailey. 1997. *Third World Political Ecology*. London: Routledge.

Butalia, Urvashi. 2000. *The Other Side of Silence: Voices from the Partition of India*. Durham, NC: Duke University Press.

Butler, Judith. 2003. "Afterword: After Loss, What Then?" In *Loss: The Politics of Mourning*, edited by David L. Eng and David Kazanjian, 467–473. Berkeley: University of California Press.

Button, Gregory. 2014. *Everyday Disasters: Rethinking Iconic Events in Cultural Perspective*. Walnut Creek: Left Coast Press.

Camargo, Alejandro, and Luisa Cortesi. 2019. "Flooding Water and Society." *Wiley Interdisciplinary Reviews: Water* 6 (5): e1374.

Campbell, Jeremy M. 2015. *Conjuring Property: Speculation and Environmental Futures in the Brazilian Amazon*. Seattle: University of Washington Press.

Campbell, John R., and Richard D. Bedford. 2022. "Climate Change and Migration: Lessons from Oceania." In *Routledge Handbook of Immigration and Refugee Studies*, edited by Anna Triandafyllidou, 379–387. London: Routledge.

Carse, Ashley. 2014. *Beyond the Big Ditch: Politics, Ecology, and Infrastructure at the Panama Canal*. Cambridge, MA: MIT Press.

Carse, Ashley, and Joshua A. Lewis. 2016. "Toward a Political Ecology of Infrastructure Standards: Or, How to Think about Ships, Waterways, Sediment, and Communities Together." *Environment and Planning A* 49 (1): 9–28. https://doi.org/10.1177/0308518X16663015.

Carse, Ashley, Townsend Middleton, Jason Cons, Jatin Dua, Gabriela Valdivia, and Elizabeth Cullen Dunn. 2020. "Chokepoints: Anthropologies of the Constricted Contemporary." *Ethnos* 88 (2): 1–11. https://doi.org/10.1080/00141844.2019.1696862.

Carson, Rachel. 1962. *Silent Spring*. Boston: Houghton Mifflin.

Castles, Stephen. 2002. *Environmental Change and Forced Migration: Making Sense of the Debate*. Geneva: United Nations High Commissioner for Refugees.

Centre for Science and Environment. 2012. *Living with Changing Climate: Impact, Vulnerability, and Adaptation Challenges in Indian Sundarbans*. New Delhi: Centre for Science and Environment.

Chakrabarti, Dilip K. 2001. *Archeological Geography of the Ganga Plain: The Lower and the Middle Ganga*. New Delhi: Permanent Black.

Chakrabarty, Dipesh. 2000. *Rethinking Working-Class History: Bengal 1890–1940*. Princeton, NJ: Princeton University Press.

———. 2001. *Provincializing Europe: Postcolonial Thought and Historical Difference*. Princeton, NJ: Princeton University Press.

———. 2009. "The Climate of History: Four Theses." *Critical Inquiry* 35:197–222.

Chandra, Uday, Geir Heierstad, and Kenneth Bo Nielsen, eds. 2015. *The Politics of Caste in West Bengal*. New Delhi: Routledge India.

Chandra, Uday, and Kenneth Bo Nielsen. 2012. "The Importance of Caste in Bengal." *Economic and Political Weekly* 47 (44): 59–61.

Chatterjee, Partha. 2006. *The Politics of the Governed: Reflections on Popular Politics in Most of the World*. New York: Columbia University Press.

———. 2011. *Lineages of Political Society: Studies in Postcolonial Democracy*. New York: Columbia University Press.

———. 2012. *The Black Hole of Empire: History of a Global Practice of Power*. Princeton, NJ: Princeton University Press.

Chattopadhyay, Gargi. 2020. "Marauders of the Sundarbans and the Role of the Island of Sagor, 1600–1800." In *An Earthly Paradise: Trade, Politics and Culture in Early Modern Bengal*, edited by Raziuddin Aquil and Tilottama Mukherjee, 83–122. London: Routledge.

Chowdhuri, Rupak De. 2018. "Villagers Fear for Survival on India's Disappearing Island." *Wider Image*, 2018. https://widerimage.reuters.com/story/villagers-fear-for-survival-on-indias-disappearing-island.

Cockburn, Alexander. 2010. "The Sinking Sundarbans." *The Independent*, January 11, 2010. http://www.independent.co.uk/environment/climate-change/the-sinking-sundarbans-1862267.html.

Colson, Elizabeth. 1971. *The Social Consequences of Resettlement: The Impact of the Kariba Resettlement upon the Gwembe Tonga*. Manchester: Manchester University Press.

———. 2003. "Forced Migration and the Anthropological Response." *Journal of Refugee Studies* 16 (1): 1–18.

Colten, Craig E. 2006. "Vulnerability and Place: Flat Land and Uneven Risk in New Orleans." *American Anthropologist* 108 (4): 731–734.

Connerton, Paul. 1989. *How Societies Remember*. Cambridge: Cambridge University Press.

———. 2008. "Seven Types of Forgetting." *Memory Studies* 1 (1): 59–71.

———. 2009. *How Modernity Forgets*. Cambridge: Cambridge University Press.

Cons, Jason. 2017. "Seepage." Theorizing the Contemporary. *Fieldsights*, October 24, 2017. https://culanth.org/fieldsights/seepage.

———. 2018. "Staging Climate Security: Resilience and Heterodystopia in the Bangladesh Borderlands." *Cultural Anthropology* 33 (2): 266–294.

———. 2021. "Ecologies of Capture in Bangladesh's Sundarbans: Predations on a Climate Frontier." *American Ethnologist* 48 (3): 245–259.

Corbridge, Stuart, Glyn Williams, Manoj Srivastava, and Rene Veron. 2002. *Seeing the State: Governance and Governmentality in India*. Cambridge: Cambridge University Press.

Crate, Susan A., and Mark Nuttall, eds. 2016. *Anthropology and Climate Change: From Actions to Transformations*. London: Routledge.

Crow, Ben, and Farhana Sultana. 2002. "Gender, Class and Access to Water: Three Cases in a Poor and Crowded Delta." *Society and Natural Resources* 15 (8): 709–724.

Cunha, Dilip Da. 2018. *The Invention of Rivers: Alexander's Eye and Ganga's Descent*. Philadelphia: University of Pennsylvania Press.

Dalby, Simon. 2005. "The Environment as Geopolitical Threat: Reading Robert Kaplan's 'Coming Anarchy.'" In *The Environment in Anthropology: A Reader in Ecology, Culture, and Sustainable Living*, edited by Nora Haenn and Richard Wilk, 118–135. New York: New York University Press.

Danda, Anamitra Anurag. 2007. "Surviving the Sundarbans: Threats and Responses. An Analytical Description of Life in an Indian Riparian Commons." PhD diss., University of Twente.

Das, Nirmalendu. n.d. "Believe It or Not: Jambudwip—An Untold Story." *DISHA Environment Newsletter*.

Das, Shaberi, and Sugata Hazra. 2020. "Trapped or Resettled: Coastal Communities in the Sundarbans Delta, India." *Forced Migration Review* (64): 15–17.

Das, Veena. 1996. *Critical Events: An Anthropological Perspective on Contemporary India*. New Delhi: Oxford University Press.

———. 2006. *Life and Words: Violence and the Descent into the Ordinary*. New Delhi: Oxford University Press.

———. 2011. "State, Citizenship, and the Urban Poor." *Citizenship Studies* 15 (3–4): 319–333.

Dasgupta, Samira, Krishna Mondal, and Krishna Basu. 2006. "Dissemination of Cultural Heritage and Impact of Pilgrim Tourism at Gangasagar Island." *Anthropologist* 8 (1): 11–15.

Dasgupta, Susmita, David Wheeler, Sunando Bandyopadhyay, Santadas Ghosh, and Utpal Roy. 2022. "Coastal Dilemma: Climate Change, Public Assistance and Population Displacement." *World Development* 150 (February): 105707. https://doi.org/10.1016/j.worlddev.2021.105707.

Davies, Thom. 2018. "Toxic Space and Time: Slow Violence, Necropolitics, and Petrochemical Pollution." *Annals of the American Association of Geographers* 108 (6): 1537–1553.

De, Barun. 1983. *West Bengal District Gazetteers: 24 Parganas.* Calcutta: Government of West Bengal.

de Laet, Marianne, and Annemarie Mol. 2000. "The Zimbabwe Bush Pump: Mechanics of a Fluid Technology." *Social Studies of Science* 30 (2): 225–263.

Descola, Philippe. 1998. *The Spears of Twilight: Life and Death in the Amazon Jungle.* New York: New Press.

De Sherbinin, Alex, Marcia Castro, Francois Gemenne, Michael M. Cernea, Susana Adamo, P. M. Fearnside, Gary Krieger, Sarah Lahmani, Anthony Oliver-Smith, and Alula Pankhurst. 2011. "Preparing for Resettlement Associated with Climate Change." *Science* 334 (6055): 456–457.

Dewan, Camelia. 2021a. "Embanking the Sundarbans: The Obfuscating Discourse of Climate Change." In *The Anthroposcene of Weather and Climate: Ethnographic Contributions to the Climate Change Debate,* edited by Paul Sillitoe, 294–321. New York: Berghahn Books.

———. 2021b. *Misreading the Bengal Delta: Climate Change, Development, and Livelihoods in Coastal Bangladesh.* Seattle: University of Washington Press.

Dhanagare, D. N. 1976. "Peasant Protest and Politics—The Tebhaga Movement in Bengal (India), 1946–47." *Journal of Peasant Studies* 3 (3): 360–378.

Dimock, Edward C. 1976. "A Theology of the Repulsive: Some Reflections on the Sitala and Other Mangals." In *Bengal: Studies in Literature, Society and History,* edited by Marvin Davis, 69–73. East Lansing: Asian Studies Center, Michigan State University.

Doocy, S., A. Dick, A. Daniels, and T. D. Kirsch. 2013. "The Human Impact of Tropical Cyclones: A Historical Review of Events 1980–2009 and Systematic Literature Review." *PLoS Currents* 5.

Doron, Assa, Richard Barz, and Barbara Nelson, eds. 2015. *An Anthology of Writings on the Ganga: Goddess and River in History, Culture, and Society.* New Delhi: Oxford University Press India.

Dove, Michael R. 2010. "The Panoptic Gaze in a Non-Western Setting: Self-Surveillance on Merapi Volcano, Central Java." *Religion* 40 (2): 121–127.

Drew, Georgina. 2017. *River Dialogues: Hindu Faith and the Political Ecology of Dams on the Sacred Ganga.* Tucson: University of Arizona Press.

D'Souza, Rohan. 2006. *Drowned and Damned: Colonial Capitalism and Flood Control in Eastern India.* New Delhi: Oxford University Press.

———. 2009. "River as Resource and Land to Own: The Great Hydraulic Transition in Eastern India." Unpublished manuscript.

DTE Staff. 2019. "Bulbul Kills in Bengal before Moving onto Bangladesh." *Down to Earth*, October 11, 2019. https://www.downtoearth.org.in/news /natural-disasters/bulbul-kills-in-bengal-before-moving-onto-bangladesh -67690.

Eaton, Richard M. 1996. *The Rise of Islam and the Bengal Frontier, 1204–1760*. Berkeley: University of California Press.

Eck, Diana L. 1998. "Gaṅgā: The Goddess in Hindu Sacred Geography." In *Devī: Goddesses of India*, edited by John Stratton Hawley and Donna M. Wulff, 137–154. New Delhi: Motilal Banarsidass.

Ecks, Stefan. 2013. *Eating Drugs: Psychopharmaceutical Pluralism in India*. New York: New York University Press.

Eden, Emily. 1866. *"Up the Country": Letters Written to Her Sister from the Upper Provinces of India*. Vol. 1. London: R. Bentley.

Elden, Stuart. 2007. "Governmentality, Calculation, Territory." *Environment and Planning D* 25 (3): 562.

El-Hinnawi, E. 1985. *Environmental Refugees*. Nairobi: United Nations High Commissioner for Refugees.

Eriksen, Thomas Hylland. 2016. *Overheating: An Anthropology of Accelerated Change*. London: Pluto Press.

Escobar, Arturo. 1999. "After Nature: Steps to an Antiessentialist Political Ecology." *Current Anthropology* 40 (1): 1–30.

Esposito, Elena. 2008. "Social Forgetting: A Systems-Theory Approach." In *Cultural Memory Studies: An International and Interdisciplinary Handbook*, edited by Astrid Erll and Ansgar Nünning, 181–189. Berlin: De Gruyter.

Ethridge, Robbie. 2006. "Bearing Witness: Assumptions, Realities, and the Otherizing of Katrina." *American Anthropologist* 108 (4): 799–813.

Farbotko, Carol. 2010. "Wishful Sinking: Disappearing Islands, Climate Refugees and Cosmopolitan Experimentation." *Asia Pacific Viewpoint* 51 (1): 47–60.

Farbotko, Carol, and Heather Lazrus. 2012. "The First Climate Refugees? Contesting Global Narratives of Climate Change in Tuvalu." *Global Environmental Change* 22 (2): 382–390.

Fardon, Richard, ed. 1990. *Localizing Strategies: Regional Traditions of Ethnographic Writing*. Edinburgh: Scottish Academic Press.

Farmer, Paul. 1996. "On Suffering and Structural Violence: A View from Below." *Daedalus* 125 (1): 261–283.

———. 2004. "An Anthropology of Structural Violence." *Current Anthropology* 45 (3): 305–325. https://doi.org/10.1086/382250.

Farrell, Clare, Alison Green, Sam Knights, and William Skeaping, eds. 2019. *This Is Not a Drill: An Extinction Rebellion Handbook*. London: Penguin.

Fergusson, James. 1863. "On Recent Changes in the Delta of the Ganges." *Quarterly Journal of the Geological Society* 19 (1–2): 321–354.

Ferrari, Fabrizio M. 2007. "'Love Me Two Times.' From Smallpox to AIDS: Contagion and Possession in the Cult of Śītalā." *Religions of South Asia* 1 (1): 81–106.

———. 2010. "Old Rituals for New Threats: Possession and Healing in the Cult of Sītalā." In *Ritual Matters: Dynamic Dimensions in Practice,* edited by Christiane Brosius and Ute Hüsken, 144–172. London: Routledge.

Finan, Timothy. 2009. "Storm Warnings: The Role of Anthropology in Adapting to Sea-Level Rise in Southwestern Bangladesh." In *Anthropology and Climate Change: From Encounters to Actions,* edited by Susan A. Crate and Mark Nuttall, 175–185. Walnut Creek, CA: Left Coast Press.

Fortun, Kim. 2000. "Remembering Bhopal, Re-figuring Liability." *Interventions* 2 (2): 187–198.

Foucault, Michel. 2012. *Discipline and Punish: The Birth of the Prison.* New York: Random House.

Franzen, Jonathan. 2019. "What If We Stopped Pretending." *New Yorker,* September 8, 2019. https://www.newyorker.com/culture/cultural-comment/what-if-we-stopped-pretending.

Freed, Ruth S., and Stanley A. Freed. 1964. "Calendars, Ceremonies, and Festivals in a North Indian Village: Necessary Calendric Information for Fieldwork." *Southwestern Journal of Anthropology* 20 (1): 67–90.

Gardner, Katy. 1991. *Songs at the River's Edge: Stories from a Bangladeshi Village.* London: Pluto Press.

———. 2012. *Discordant Development: Global Capitalism and the Struggle for Connection in Bangladesh.* London: Pluto Press.

Garrison, Tom. 2011. *Essentials of Oceanography.* Belmont: Brooks/Cole.

Geertz, Clifford. 1972. "The Wet and the Dry: Traditional Irrigation in Bali and Morocco." *Human Ecology* 1 (1): 23–39.

Gesing, Friederike. 2016. *Working with Nature in Aotearoa New Zealand: An Ethnography of Coastal Protection.* Bielefeld: transcript.

Ghertner, D. Asher. 2011. "Green Evictions: Environmental Discourses of a Slum-Free Delhi." In *Global Political Ecology,* edited by Richard Peet, Paul Robbins, and Michael J. Watts, 145–165. London: Routledge.

———. 2015. *Rule by Aesthetics: World-Class City Making in Delhi.* New York: Oxford University Press.

Ghosh, Aditya. 2017. *Sustainability Conflicts in Coastal India: Hazards, Changing Climate and Development Discourses in the Sundarbans.* Dordrecht: Springer.

Ghosh, Amitav. 2004. *The Hungry Tide.* London: HarperCollins.

———. 2019. *Gun Island.* London: John Murray.

Ghosh, Tuhin, Gupinath Bhandari, and Sugata Hazra. 2003. "Application of 'Bio-Engineering' Technique to Protect Ghoramara Island (Bay of Bengal) from Severe Erosion." *Journal of Coastal Research* 9 (2): 171–178.

Ghosh, Upasona, Shibaji Bose, and Rittika Bramhachari. 2018. *Living on the Edge: Climate Change and Uncertainty in the Indian Sundarbans.* STEPS Working Paper 101. Brighton: STEPS Centre.

Gibson, Hannah, and Sita Venkateswar. 2015. "Anthropological Engagement with the Anthropocene: A Critical Review." *Environment and Society: Advances in Research* 6 (1): 5–27.

Giri, Aswini Kumar. 2007. *Hijlir Pirbaba Masnad-I-Ala* [The throne of Hijli's Pirbaba]. Kanthi: Arabindra Granthasangraha.

Glantz, Michael. 1999. "Sustainable Development and Creeping Environmental Problems in the Aral Sea Region." In *Creeping Environmental Problems and Sustainable Development in the Aral Sea Basin,* edited by Michael Glantz, 1–25. Cambridge: Cambridge University Press.

Gold, Ann Grodzins, and Bhoju Ram Gujar. 2002. *In the Time of Trees and Sorrows: Nature, Power, and Memory in Rajasthan.* Durham, NC: Duke University Press.

Goldsmith, Michael. 2015. "The Big Smallness of Tuvalu." *Global Environment* 8 (1): 134–151.

Gordillo, Gastón R. 2014. *Rubble: The Afterlife of Destruction.* Durham, NC: Duke University Press.

Graham, Maria. 1812. *Journal of a Residence in India.* Edinburgh: Archibald Constable.

Graham, Steve, and Simon Marvin. 2002. *Splintering Urbanism: Networked Infrastructures, Technological Mobilities and the Urban Condition.* London: Taylor and Francis.

Gray, Peter O., and Kendrick Oliver, eds. 2004. *The Memory of Catastrophe.* Manchester: Manchester University Press.

Greenough, Paul. 1998. "Hunter's Drowned Land: An Environmental Fantasy of the Victorian Sundarbans." In *Nature and the Orient: Environmental History of South and Southeast Asia,* edited by Richard H. Grove, Vinita Damodaran, and Satpal Sangwan, 237–272. New Delhi: Oxford University Press.

———. 2003. "Pathogens, Pugmarks, and Political 'Emergency': The 1970's South Asian Debate on Nature." In *Nature in the Global South: Environmental Projects in South and Southeast Asia,* edited by Paul Greenough and Anna L. Tsing, 201–230. Durham, NC: Duke University Press.

Greiner, Clemens, and Patrick Sakdapolrak. 2013. "Translocality: Concepts, Applications and Emerging Research Perspectives." *Geography Compass* 7 (5): 373–384.

Gupta, Akhil. 2012. *Red Tape: Bureaucracy, Structural Violence, and Poverty in India*. Durham, NC: Duke University Press.

Gupta, Charu, and Mukul Sharma. 2009. *Contested Coastlines: Fisherfolk, Nations and Borders in South Asia*. New Delhi: Routledge.

Gupta, Joydeep. 2018. "Rising Sea Swamps Island along Bengal Coast." *Third Pole* (blog). January 15, 2018. https://www.thethirdpole.net/2018/01/15 /rising-sea-swamps-island-along-bengal-coast/.

Haberman, David L. 2006. *River of Love in an Age of Pollution: The Yamuna River of Northern India*. Berkeley: University of California Press.

———. 2013. *People Trees: Worship of Trees in Northern India*. Oxford: Oxford University Press.

Halbwachs, Maurice. 1992. *On Collective Memory*. Chicago: University of Chicago Press.

Hallmann, Caspar A., Martin Sorg, Eelke Jongejans, Henk Siepel, Nick Hofland, Heinz Schwan, Werner Stenmans, Andreas Müller, Hubert Sumser, and Thomas Hörren. 2017. "More than 75 Percent Decline over 27 Years in Total Flying Insect Biomass in Protected Areas." *PloS One* 12 (10): e0185809.

Hamilton, Walter. 1815. *The East India Gazetteer*. London: John Murray.

Haque, C. E. 1997. *Hazards in a Fickle Environment: Bangladesh*. Dordrecht: Springer.

Haraway, Donna, Noboru Ishikawa, Scott F. Gilbert, Kenneth Olwig, Anna L. Tsing, and Nils Bubandt. 2016. "Anthropologists Are Talking—About the Anthropocene." *Ethnos* 81 (3): 535–564. https://doi.org/10.1080/00141844 .2015.1105838.

Harms, Arne. 2015. "Leaving Lohachara: On Circuits of Displacement and Emplacement in the Indian Ganges Delta." *Global Environment* 8 (1): 62–85.

———. 2017. "Citizenship at Sea: Environmental Displacement and State Relations in the Indian Sundarbans." *Economic and Political Weekly* 52 (33): 69–76.

———. 2018. "Filming Sea-level Rise: Media Encounters and Memory Work in the Indian Sundarbans." *Journal of the Royal Anthropological Institute* 24 (3): 475–492. https://doi.org/10.1111/1467-9655.12856.

———. 2019. *Infrastrukturen* [Infrastructures]. Berlin: De Gruyter.

———. Forthcoming. "Weathering the Sea: The Coast as Crumple Zone" In *Coastal Futures*, edited by Lukas Ley and Arne Harms. Toronto: University of Toronto Press.

Harms, Arne, and Oliver Powalla. 2014. "India—The Long March to a Climate Movement." In *Routledge Handbook of the Climate Change Movement*, edited by Matthias Dietz and Heiko Garrelts, 179–193. London: Routledge.

Hartmann, Betsy. 2001. "Will the Circle Be Unbroken? A Critique of the Project on Environment, Population, and Security." In *Violent Environments*, edited by Nancy Lee Peluso and Michael Watts, 39–62. Ithaca, NY: Cornell University Press.

———. 2010. "Rethinking Climate Refugees and Climate Conflict: Rhetoric, Reality and the Politics of Policy Discourse." *Journal of International Development* 22:233–246.

Hartmann, Betsy, and James K. Boyce. 1983. *A Quiet Violence: View from a Bangladesh Village*. London: Zed Press.

Hastrup, Frida. 2011. *Weathering the World: Recovery in the Wake of the Tsunami in a Tamil Fishing Village*. New York: Berghahn.

Hazra, Sugata. 2012. "Climate Change Policy Paper II: Climate Change Adaptation in Coastal Region of West Bengal." New Delhi: WWF India. http://awsassets.wwfindia.org/downloads/climate_change_adaptation_in_coastal_region_of_west_bengal.pdf.

Hoffman, Susana M. 1999. "The Worst of Times, the Best of Times: Toward a Model of Cultural Response to Disaster." In *The Angry Earth: Disaster in Anthropological Perspective*, edited by Anthony Oliver-Smith and Susanna M. Hoffman, 141–161. New York: Routledge.

———. 2002. "The Monster and the Mother: The Symbolism of Disaster." In *Catastrophe and Culture: The Anthropology of Disaster*, edited by Susana M. Hoffman and Anthony Oliver-Smith, 113–142. Santa Fe, NM: School of American Research Press.

Howe, Cymene, Jessica Lockrem, Hannah Appel, Edward Hackett, Dominic Boyer, Randal Hall, Matthew Schneider-Mayerson, Albert Pope, Akhil Gupta, and Elizabeth Rodwell. 2016. "Paradoxical Infrastructures: Ruins, Retrofit, and Risk." *Science, Technology, and Human Values* 41 (3): 547–565.

Huggins, William. 1824. *Sketches in India: Treating on Subjects Connected with the Government; Civil and Military Establishments; Characters of the European, and Customs of the Native Inhabitants*. London: Letts.

Hull, Matthew S. 2012. *Government of Paper: The Materiality of Bureaucracy in Urban Pakistan*. Berkeley: University of California Press.

Hulme, Mike. 2011. "Reducing the Future to Climate: A Story of Climate Determinism and Reductionism." *Osiris* 26 (1): 245–266.

Hunter, Lori M. 2005. "Migration and Environmental Hazards." *Population and Environment* 26 (4): 273–302.

Hunter, Lori M., Jessie K. Luna, and Rachel M. Norton. 2015. "Environmental Dimensions of Migration." *Annual Review of Sociology* 41:377–397.

Hunter, William Wilson. 1875. *A Statistical Account of Bengal Volume I: Districts of the 24 Parganas and Sundarbans.* London: Trübner.

Hutnyk, John. 1996. *The Rumour of Calcutta: Tourism, Charity and the Poverty of Representation.* London: Zed Books.

Hutton, David, and C. Emdad Haque. 2003. "Patterns of Coping and Adaptation Among Erosion-Induced Displacees in Bangladesh: Implications for Hazard Analysis and Mitigation." *Natural Hazards* 29:405–421.

———. 2004. "Human Vulnerability, Dislocation and Resettlement: Adaptation Processes of River-Bank Erosion-Induced Displacees in Bangladesh." *Disasters* 28 (1): 41–62.

IDMC. 2021. *Internal Displacement in a Changing Climate.* Geneva: Internal Displacement Monitoring Center.

Ingold, Tim. 2000. *The Perception of the Environment: Essays on Livelihood, Dwelling and Skill.* London: Routledge.

Iqbal, Iftekhar. 2010. *The Bengal Delta: Ecology, State and Social Change, 1840–1943.* Basingstoke: Palgrave Macmillan.

Iqbal, Showkat. 2010. "Flood and Erosion Induced Population Displacements: A Socio-Economic Case Study in the Gangetic Riverine Tract at Malda District, West Bengal, India." *Journal of Human Ecology* 30 (3): 201–211.

Islam, M. Rezaul, and Mehedi Hasan. 2016. "Climate-Induced Human Displacement: A Case Study of Cyclone Aila in the South-West Coastal Region of Bangladesh." *Natural Hazards* 81 (2): 1051–1071. https://doi.org/10.1007/s11069-015-2119-6.

Islam, Mohammad Rezwanul, Syeda Fahliza Begum, Yasushi Yamaguchi, and Katsuro Ogawa. 1999. "The Ganges and Brahmaputra Rivers in Bangladesh: Basin Denudation and Sedimentation." *Hydrological Processes* 13 (17): 2907–2923.

Jacobsen, Knut A. 2008. *Kapila, Founder of Sāṃkhya and Avatāra of Viṣṇu: With a Translation of Kapilāsurisaṃvāda.* New Delhi: Munshiram Manoharlal.

———. 2012. *Pilgrimage in the Hindu Tradition: Salvific Space.* London: Routledge.

Jalais, Annu. 2005. "Dwelling on Morichjhanpi: When Tigers Became 'Citizens', Refugees 'Tiger-Food.'" *Economic and Political Weekly* 40 (17): 1757–1762.

———. 2008. "Unmasking the Cosmopolitan Tiger." *Nature and Culture* 3 (1): 25–40.

———. 2010a. "Braving Crocodiles with Kali: Being a Prawn Seed Collector and a Modern Woman in the 21st Century Sundarbans." *Socio-Legal Revue* 6 (1): 1–23.

———. 2010b. *Forest of Tigers: People, Politics and Environment in the Sundarbans.* New Delhi: Routledge.

Jalais, Annu, and Amites Mukhopadhyay. 2020. "Of Pandemics and Storms in the Sundarbans." In "Intersecting Crises," edited by Calynn Dowler. American Ethnologist Website, October 12, 2020. https://americanethnologist .org/online-content/collections/intersecting-crises/of-pandemics-and -storms-in-the-sundarbans/.

Jay, Martin. 1993. *Downcast Eyes: The Denigration of Vision in Twentieth-Century French Thought.* Berkeley: University of California Press.

Johnson, Andrew Alan. 2019. "The River Grew Tired of Us: Spectral Flows along the Mekong River." *HAU: Journal of Ethnographic Theory* 9 (2): 390–404.

Kane, Stephanie. 2012. *Where Rivers Meet the Sea: The Political Ecology of Water.* Philadelphia: Temple University Press.

Kanjilal, Tushar. 2000. *Who Killed the Sundarbans?* Calcutta: Tagore Society for Rural Development.

Kankara, R. S., M. V. Ramana Murthy, and M. Rajeevan. 2018. *National Assessment of Shoreline Changes along Indian Coast: Status Report for 26 Years (1990–2016).* Chennai: Ministry of Earth Sciences National Centre for Coastal Research.

Kapila, Kriti. 2022. *Nullius: The Anthropology of Ownership, Sovereignty, and the Law in India.* Chicago: HAU Books.

Kaul, Suvir, ed. 2001. *The Partitions of Memory: The Afterlife of the Division of India.* Bloomington: Indiana University Press.

Kawa, Nicholas C. 2016. *Amazonia in the Anthropocene: People, Soils, Plants, Forests.* Austin: University of Texas Press.

Keller, Kirsten. 2023. "Mussels and Megaprojects: Landscape Structure and Structural Inequality at Jakarta's Coast." *Social Anthropology/Anthropologie Sociale* 31 (4): 76–94.

Kelman, Ilan. 2019. "Imaginary Numbers of Climate Change Migrants?" *Social Sciences* 8 (5): 131.

Kemmer, Laura, and AbdouMaliq Simone. 2021. "Standing by the Promise: Acts of Anticipation in Rio and Jakarta." *Environment and Planning D: Society and Space* 39 (4): 573–589.

Khan, Naveeda. 2016. "Living Paradox in Riverine Bangladesh: Whiteheadian Perspectives on Ganga Devi and Khwaja Khijir." *Anthropologica*, 179–192.

———. 2022. *River Life and the Upspring of Nature.* Durham, NC: Duke University Press.

Kingsbury, Benjamin. 2015. "An Imperial Disaster: The Bengal Cyclone of 1876." PhD diss., Victoria University of Wellington.

Kleinman, Arthur, Veena Das, and Margaret Lock. 1996. "Introduction." *Daedalus* 125 (1): xi–xx.

Klepp, Silja. 2018. "Climate Change and Migration." In *Oxford Research Encyclopedia of Climate Science,* edited by Matthew Nisbet. Oxford: Oxford University Press.

Kolbert, Elizabeth. 2014. *The Sixth Extinction: An Unnatural History.* New York: Henry Holt.

Koslov, Liz. 2016. "The Case for Retreat." *Public Culture* 28 (2): 359–387. https://doi.org/10.1215/08992363-3427487.

Krause, Franz. 2017. "Towards an Amphibious Anthropology of Delta Life." *Human Ecology* 45:403–408.

Krishnamurthy, Rohini. 2023. "Mangroves Protect Coastal Communities, Accountability in Governance Is Need of the Hour: Experts." *Down to Earth,* February 7, 2023. https://www.downtoearth.org.in/news/wildlife -biodiversity/mangroves-protect-coastal-communities-accountability-in -governance-is-need-of-the-hour-experts-87524.

Kuletz, Valerie. 2001. "Invisible Spaces, Violent Places: Cold War Nuclear and Militarized Landscapes." In *Violent Environments,* edited by Nancy Lee Peluso and Michael Watts, 237–260. Ithaca, NY: Cornell University Press.

Kulp, Scott A., and Benjamin H. Strauss. 2019. "New Elevation Data Triple Estimates of Global Vulnerability to Sea-Level Rise and Coastal Flooding." *Nature Communications* 10 (1): 1–12. https://doi.org/10.1038/s41467-019 -12808-z.

Lahiri, Anil Chandra. 1936. *Final Report on the Survey and Settlement Operations in the District of 24 Parganas, 1924–33.* Calcutta: Bengal Government Press.

Lahiri-Dutt, Kuntala, and Gopa Samanta. 2007. "'Like the Drifting Grains of Sand': Vulnerability, Security and Adjustment by Communities in the Charlands of the Damodar Delta." *South Asia: Journal of South Asia Studies* 32 (2): 320–357.

———. 2013. *Dancing with the River: People and Life on the Chars of South Asia.* New Haven, CT: Yale University Press.

Lambek, Michael. 1996. "The Past Imperfect: Remembering as Moral Practice." In *Tense Past: Cultural Essays in Trauma and Memory,* edited by Paul Antze and Michael Lambek, 235–254. London: Routledge.

Larkin, Brian. 2013. "The Politics and Poetics of Infrastructure." *Annual Review of Anthropology* 42:327–343.

Latour, Bruno. 1993. *We Have Never Been Modern.* Cambridge, MA: Harvard University Press.

———. 2018. *Down to Earth: Politics in the New Climatic Regime.* London: John Wiley and Sons.

Lauer, Matthew, Simon Albert, Shankar Aswani, Benjamin S. Halpern, Luke Campanella, and Douglas La Rose. 2013. "Globalization, Pacific Islands, and the Paradox of Resilience." *Global Environmental Change* 23 (1): 40–50.

Lein, Haakon. 2009. "The Poorest and Most Vulnerable? On Hazards, Livelihoods and Labelling of Riverine Communities in Bangladesh." *Singapore Journal of Tropical Geography* 30 (1): 98–113.

Lemke, Thomas. 2015. "New Materialisms: Foucault and the 'Government of Things.'" *Theory, Culture and Society* 32 (4): 3–25.

Lévi-Strauss, Claude. 2012. *Tristes Tropiques.* London: Penguin.

Ley, Lukas. 2018. "On the Margins of the Hydrosocial: Quasi-Events along a Stagnant River." *Geoforum* 131:234–242.

———. 2021. *Building on Borrowed Time: Rising Seas and Failing Infrastructure in Semarang.* Minneapolis: University of Minnesota Press.

Li, Tania Murray. 2014. *Land's End: Capitalist Relations on an Indigenous Frontier.* Durham, NC: Duke University Press.

Liboiron, Max, Manuel Tironi, and Nerea Calvillo. 2018. "Toxic Politics: Acting in a Permanently Polluted World." *Social Studies of Science* 48 (3): 331–349.

Lieten, G. K. 1996. "Land Reforms at Centre Stage: The Evidence on West Bengal." *Development and Change* 27 (1): 111–130.

Lübken, Uwe. 2012. "Chasing a Ghost? Environmental Change and Migration in History." *Global Environment* 9:1–17.

Luig, Ute. 2012. "Negotiating Disaster: An Overview." In *Negotiating Disasters: Politics, Representation, Meanings,* edited by Ute Luig, 3–26. Frankfurt: Peter Lang.

Lutgendorf, Philip. 2006. "A Ramayana on Air: 'All in the (Raghu) Family,' A Video Epic in Cultural Context." In *The Life of Hinduism,* edited by John Stratton Hawley and Vasudha Nar, 140–157. Berkeley: University of California Press.

Maiti, Jagannath. 1994. *Sagordvipke Janun* [Know Sagar Island]. Kalibajar: Mokshda Pustokaloy.

———. 2008. *Sagardviper Palligram* [The land of Sagar Island]. Mansadvip: Sarbasattba Samraksita.

Maldonado, Julie K. 2018. *Seeking Justice in an Energy Sacrifice Zone: Standing on Vanishing Land in Coastal Louisiana*. London: Routledge.

Malkki, Liisa H. 1995. *Purity and Exile: Violence, Memory, and National Cosmology Among Hutu Refugees in Tanzania*. Chicago: University of Chicago Press.

Mallick, Ross. 1993. *Development Policy of a Communist Government: West Bengal Since 1977*. Cambridge: Cambridge University Press.

———. 1999. "Refugee Resettlement in Forest Reserves: West Bengal Policy Reversal and the Marichjhapi Massacre." *Journal of Asian Studies* 58 (1): 104–125.

Manning, Erin. 2016. *The Minor Gesture*. Durham, NC: Duke University Press.

Marchezini, Victor. 2015. "The Biopolitics of Disaster: Power, Discourses, and Practices." *Human Organization* 74 (4): 362–371.

Marino, Elizabeth. 2013. "Environmental Migration: The Future of Anthropology in Social Vulnerability, Disaster and Discourse." In *Environmental Anthropology: Future Directions*, edited by Helen Kopnina and Eleanor Shoreman-Ouimet, 188–203. New York: Routledge.

———. 2015. *Fierce Climate, Sacred Ground: An Ethnography of Climate Change in Shishmaref, Alaska*. Fairbanks: University of Alaska Press.

Masco, Joseph. 2015. "The Age of Fallout." *History of the Present* 5 (2): 137–168.

Massey, Deeptima. 2009. "Seeking Informal Social Protection: Migrant Households in Rural West Bengal." In *Circular Migration and Multilocational Livelihood Strategies in Rural India*, edited by Priya Deshingkar and John Farrington, 278–295. New Delhi: Oxford University Press.

Mathur, Hari Mohan. 2015. "Climate Change and Displacement: Learning from Resettlement in the Development Context." *Social Change* 45 (1): 118–130. https://doi.org/10.1177/0049085714561939.

Mathur, Nayanika. 2015. *Paper Tiger: Law, Bureaucracy and the Developmental State in Himalayan India*. Cambridge: Cambridge University Press.

McDermott, Rachel Fell. 2013. *Revelry, Rivalry, and Longing for the Goddesses of Bengal: The Fortunes of Hindu Festivals*. New York: Columbia University Press.

McLeman, Robert, and François Gemenne. 2018. "Environmental Migration Research: Evolution and Current State of the Science." In *Routledge Handbook of Environmental Displacement and Migration*, edited by Robert McLeman and François Gemenne, 3–16. London: Routledge.

Merry, Sally Engle. 2016. *The Seductions of Quantification: Measuring Human Rights, Gender Violence, and Sex Trafficking*. Chicago: University of Chicago Press.

Meyer, Morgan. 2012. "Placing and Tracing Absence: A Material Culture of the Immaterial." *Journal of Material Culture* 17 (1): 103–110.

Michaels, Axel. 2006. *Der Hinduismus: Geschichte und Gegenwart* [Hinduism: Past and present]. Frankfurt: Beck.

Miller, Michelle Ann, and Tim Bunnell. 2011. "Post-Disaster Urban Renewal: Memories of Trauma and Transformation in an Indonesian City." ARI Working Paper 154.

Mol, Annemarie. 2002. *The Body Multiple: Ontology in Medical Practice*. Durham, NC: Duke University Press.

Molesky, Mark. 2015. *This Gulf of Fire: The Destruction of Lisbon, or Apocalypse in the Age of Science and Reason*. New York: Vintage.

Moore, Amelia. 2016. "Anthropocene Anthropology: Reconceptualizing Contemporary Global Change." *Journal of the Royal Anthropological Institute* 22 (1): 27–46. https://doi.org/10.1111/1467-9655.12332.

Morrissey, James. 2012. "Rethinking the 'Debate on Environmental Refugees': From 'Maximilists and Minimalists' to 'Proponents and Critics.'" *Journal of Political Ecology* 19:36–49.

Mortreux, Colette, Ricardo Safra de Campos, W. Neil Adger, Tuhin Ghosh, Shouvik Das, Helen Adams, and Sugata Hazra. 2018. "Political Economy of Planned Relocation: A Model of Action and Inaction in Government Responses." *Global Environmental Change* 50 (May): 123–132. https://doi.org/10.1016/j.gloenvcha.2018.03.008.

Mukerjee, Radhakamal. 1938. *The Changing Face of Bengal: A Study in Riverine Economy*. Calcutta: University of Calcutta.

Mukherjee, Nilmani. 1968. *The Port of Calcutta: A Short History*. Calcutta: Commissioners for the Port of Calcutta.

Mukherjee, Upamanyu Pablo. 2013. *Natural Disasters and Victorian Empire: Famines, Fevers and the Literary Cultures of South Asia*. Dordrecht: Springer.

Mukhopadhyay, Amites. 2009. *Cyclone Aila and the Sundarbans: An Enquiry into the Disaster and Politics of Aid and Relief*. Kolkata: Mahanirban Calcutta Research Group.

———. 2016. *Living with Disasters: Communities and Development in the Indian Sundarbans*. Delhi: Cambridge University Press.

Murphy, Michelle. 2015. "Chemical Infrastructures of the St Clair River." In *Toxicants, Health and Regulation since 1945*, edited by Soraya Boudiya and Natalie Jas, 103–115. London: Routledge.

Naipaul, Vidiadhar Surajprasad. 1987. *The Enigma of Arrival*. New York: Knopf.

Nancy, Jean-Luc. 2014. *After Fukushima: The Equivalence of Catastrophes.* Translated by Charlotte Mandell. New York: Fordham University Press.

Nandy, Sreetapa, and Sunando Bandyopadhyay. 2011. "Trend of Sea Level Change in the Hugli Estuary, India." *Indian Journal of Geo-Marine Sciences* 40 (6): 802–812.

Narayanan, Nayantara. 2015. "The Shrinking Islands of the World's Largest Mangroves Have Triggered a Refugee Crisis." *Quartz India,* March 18, 2015. https://qz.com/india/364896/the-shrinking-islands-of-the-worlds -largest-mangroves-have-triggered-a-refugee-crisis/.

Nath, Sankar Kumar, Debasis Roy, and Kiran Kumar Singh Thingbaijam. 2008. "Disaster Mitigation and Management for West Bengal, India: An Appraisal." *Current Science* 94 (7): 858–864.

Neiman, Susan. 2002. *Evil in Modern Thought: An Alternative History of Philosophy.* Princeton, NJ: Princeton University Press.

Neumann, Barbara, Athanasios T. Vafeidis, Juliane Zimmermann, and Robert J. Nicholls. 2015. "Future Coastal Population Growth and Exposure to Sea-Level Rise and Coastal Flooding—A Global Assessment." *PloS One* 10 (3): e0118571.

Nicholas, Ralph Wallace. 1981. "The Goddess Śītalā and Epidemic Smallpox in Bengal." *Journal of Asian Studies* 41 (1): 21–44.

———. 2003. *Fruits of Worship: Practical Religion in Bengal.* New Delhi: Chronicle Books.

Nielsen, Kenneth Bo. 2018. *Land Dispossession and Everyday Politics in Rural Eastern India.* New York: Anthem Press.

Nixon, Rob. 2011. *Slow Violence and the Environmentalism of the Poor.* Cambridge, MA: Harvard University Press.

Niyogi, Subhro. 2009. "Hungry Tide, Homeless People." *Times of India,* December 5, 2009. https://timesofindia.indiatimes.com/india/hungry-tide-homeless -people/articleshow/5304605.cms.

O'Donnell, Tayanah. 2022. "Managed Retreat and Planned Retreat: A Systematic Literature Review." *Philosophical Transactions of the Royal Society B* 377 (1854): 20210129.

Oliver-Smith, Anthony. 1996. "Anthropological Research on Hazards and Disasters." *Annual Review of Anthropology* 25:303–328.

———. 1999a. "The Brotherhood of Pain: Theoretical and Applied Perspectives on Post-Disaster Solidarity." In *The Angry Earth: Disaster in Anthropological Perspective,* edited by Anthony Oliver-Smith and Susanna M. Hoffman, 156–172. New York: Routledge.

———. 1999b. "'What Is a Disaster?': Anthropological Perspectives on a Persistent Question." In *The Angry Earth: Disasters in Anthropological*

Perspective, edited by Anthony Oliver-Smith and Susanna M. Hoffman, 18–34. New York: Routledge.

———. 2009. "Climate Change and Population Displacement: Disasters and Diasporas in the Twenty-First Century." In *Anthropology and Climate Change: From Encounters to Actions,* edited by Susan A. Crate and Mark Nuttal, 116–138. Walnut Creek, CA: Left Coast Press.

———. 2011. *Defying Displacement: Grassroots Resistance and the Critique of Development.* Austin: University of Texas Press.

———. 2012. "Debating Environmental Migration: Society, Nature and Population Displacement in Climate Change." *Journal of International Development* 24 (8): 1058–1070. https://doi.org/10.1002/jid.2887.

———. 2019. "Resettlement for Disaster Risk Reduction: Global Knowledge, Local Application." In *Disaster upon Disaster: Exploring the Gap between Knowledge, Policy and Practice,* edited by Susanna M. Hoffman and Roberto E. Barrios, 198–217. New York: Berghahn Books.

O'Malley, L. S. S. 1914. *24-Parganas.* Bengal District Gazetteers. Calcutta: Bengal Secretariat Book Depot.

Ortner, Sherry. 1995. "Resistance and the Problem of Ethnographic Refusal." *Comparative Studies in Society and History* 37 (1): 173–193.

———. 2016. "Dark Anthropology and Its Others: Theory since the Eighties." *HAU: Journal of Ethnographic Theory* 6 (1): 47–73.

Padma, T. V. 2019. "Coping with Storm Surges in the Sundarbans." *Third Pole* (blog). February 1, 2019. https://www.thethirdpole.net/en/2019/02/01/coping-with-storm-surges-in-the-sundarbans/.

Pandey, Gyanendra. 2006. *Routine Violence: Nations, Fragments, Histories.* New Delhi: Permanent Black.

Pandian, Anand. 2019. *A Possible Anthropology: Methods for Uneasy Times.* Durham, NC: Duke University Press.

Paprocki, Kasia. 2019. "All That Is Solid Melts into the Bay: Anticipatory Ruination and Climate Change Adaptation." *Antipode* 51 (1): 295–315.

Pargiter, Frederick Eden. 1934. *A Revenue History of the Sundarbans, from 1765 to 1870.* Alipore: Bengal Government Press.

Parry, Jonathan P. 2008. "The Sacrifices of Modernity in a Soviet-Built Steel Town in Central India." In *On the Margins of Religion,* edited by Frances Pine and Joao de Pina-Cabral, 233–262. London: Berghahn.

Patnaik, Prabhat. 2007. "In the Aftermath of Nandigram." *Economic and Political Weekly* 42 (21): 1893–1895.

Paul, Bimal Kanti. 2010. "Human Injuries Caused by Bangladesh's Cyclone Sidr: An Empirical Study." *Natural Hazards* 54 (2): 483–495.

Peet, Richard, Paul Robbins, and Michael Watts. 2011. "Global Nature." In *Global Political Ecology*, edited by Richard Peet, Paul Robbins, and Michael Watts, 1–47. London: Routledge.

Pelling, Mark. 2003. *The Vulnerability of Cities: Natural Disasters and Social Resilience*. London: Earthscan.

Pfister, Christian. 2009. "Die 'Katastrophenlücke' Des 20. Jahrhunderts Und Der Verlust Traditionalen Risikobewusstseins" [The "disaster gap" of the twentieth century and the loss of traditional risk awareness]. *GAIA* 18 (3): 239–246.

Piguet, Etienne, and Frank Laczko. 2014. *People on the Move in a Changing Climate: The Regional Impact of Environmental Change on Migration*. Dordrecht: Springer.

Piguet, Etienne, Antoine Pécoud, and Paul de Guchteneire. 2011. "Migration and Climate Change: An Overview." *Refugee Survey Quarterly* 30 (3): 1–23.

Pilkey, Orrin H., and Keith C. Pilkey. 2019. *Sea Level Rise: A Slow Tsunami on America's Shores*. Durham, NC: Duke University Press.

Pilkey, Orrin H., Linda Pilkey-Jarvis, and Keith C. Pilkey. 2016. *Retreat from a Rising Sea: Hard Choices in an Age of Climate Change*. New York: Columbia University Press.

Povinelli, Elizabeth A. 2011. *Economies of Abandonment: Social Belonging and Endurance in Late Liberalism*. Durham, NC: Duke University Press.

Quarantelli, Enrico L. 2005. "A Social Science Research Agenda for the Disasters of the 21st Century: Theoretical, Methodological and Empirical Issues and Their Professional Implementation." In *What Is a Disaster? New Answers to Old Questions*, edited by Ronald W. Perry and Enrico L. Quarantelli, 325–396. Philadelphia: Xlibris.

Rahim, Kazi M. B., Malay Mukhopadhyay, and Debashis Sarkar, eds. 2008. *River Bank Erosion and Land Loss*. Kolkata: Visva-Bharati.

Rajan, S. Ravi. 2001. "Toward a Metaphysics of Environmental Violence: The Case of the Bhopal Gas Disaster." In *Violent Environments*, edited by Nancy Lee Peluso and Michael Watts, 380–398. Ithaca, NY: Cornell University Press.

Ramakrishna Mission Lokashiksha Parishad. 2009. *The Role of Embankments in the Livelihoods of the Local People in the Sundarbans of West Bengal, India: A Study on "Socio-Economic Context and Consequences."* Kolkata: Ramakrishna Mission Ashrama.

Ramaswamy, Sumathi. 2005. *Fabulous Geographies, Catastrophic Histories: The Lost Land of Lemuria*. Delhi: Permanent Black.

Rao, Ursula. 2018. "Biometric Bodies, or How to Make Electronic Fingerprinting Work in India." *Body and Society* 24 (3): 68–94.

Rasid, Harun, and Bimal Kanti Paul. 1987. "Flood Problems in Bangladesh: Is There an Indigenous Solution?" *Environmental Management* 11 (2): 155–173.

Ray, Achintyarup. 2009. "Vanishing Islands: Blame on KoPT." *Times of India*, April 3, 2009. http://timesofindia.indiatimes.com/city/kolkata/Vanishing-islands-Blame-on-KoPT/articleshow/4352474.cms.

Ray, Raka, and Seemin Qayum. 2009. *Cultures of Servitude: Modernity, Domesticity, and Class in India*. Stanford, CA: Stanford University Press.

Ray, Sarah Jaquette. 2020. *A Field Guide to Climate Anxiety: How to Keep Your Cool on a Warming Planet*. Oakland: University of California Press.

Raychaudhuri, Bikash. 1980. *The Moon and the Net: Study of a Transient Community of Fishermen at Jambudwip*. Calcutta: Anthropological Survey of India.

Read, Peter. 1996. *Returning to Nothing: The Meaning of Lost Places*. Cambridge: Cambridge University Press.

Redfield, Peter. 2012. "Bioexpectations: Life Technologies as Humanitarian Goods." *Public Culture* 24 (1): 157–184.

Revet, Sandrine. 2011. "Remembering La Tragedia: Commemorations of the 1999 Floods in Venezuela." In *Grassroots Memorials: The Politics of Memorializing Traumatic Death*, edited by Peter Jan Margry and Cristina Sanchez-Carretero, 208–228. New York: Berghahn.

Richards, John F., and Elizabeth P. Flint. 1990. "Long-Term Transformations in the Sundarbans Wetlands Forests of Bengal." *Agriculture and Human Values* 7 (2): 17–33.

Right to Food Campaign. 2007. *Employment Guarantee Act: A Primer*. Delhi: Right to Food.

Riles, Annelise. 2013. "Market Collaboration: Finance, Culture, and Ethnography after Neoliberalism." *American Anthropologist* 115 (4): 555–569.

Rodrigues, Hillary Peter. 2003. *Ritual Worship of the Great Goddess: The Liturgy of the Durga Puja with Interpretations*. New York: Stated University of New York Press.

Roitman, Janet. 2013. *Anti-Crisis*. Durham, NC: Duke University Press.

Roy, Ananya. 2003. *City Requiem, Calcutta: Gender and the Politics of Poverty*. Minneapolis: University of Minnesota Press.

Roy, Beth. 1994. *Some Trouble with Cows: Making Sense of Social Conflict*. Berkeley: University of California Press.

Rudiak-Gould, Peter. 2012. "Promiscuous Corroboration and Climate Change Translation: A Case Study from the Marshall Islands." *Global Environmental Change* 22 (1): 46–54.

———. 2013. *Climate Change and Tradition in a Small Island State: The Rising Tide*. London: Routledge.

Rudra, Kalyan. 2007. "The Proposed Chemical Hub in Nayachar: Some Argumentative Issues of Concern." *Counterviews Webzine* 1 (1): 3–7.

Rushdie, Salman. 1995. *Midnight's Children*. London: Vintage.

Ruud, Arild Engelsen. 2000. "Talking Dirty about Politics: A View from a Bengali Village." In *The Everyday State and Society in Modern India*, edited by Véronique Bénéï, Christopher John Fuller, and Veronique Benei, 115–136. London: Hurst.

Rycroft, Daniel J. 2006. *Representing Rebellion: Visual Aspects of Counter-Insurgency in Colonial India*. New Delhi: Oxford University Press.

Saha, Pradip, dir. 2009. *Mean Sea Level*. DVD. New Delhi: Centre for Science and Environment.

Sahlins, Marshall. 1987. *Islands of History*. Chicago: University of Chicago Press.

Said, Edward W. 2014. *Orientalism*. New York: Random House.

Sainath, P. 1996. *Everybody Loves a Good Drought: Stories from India's Poorest Districts*. New Delhi: Penguin Books India.

Samling, Clare Lizamit, Ashish Ghosh, and Sugata Hazra. 2015. *Resettlement and Rehabilitation: Indian Scenario*. DECCMA Working Paper, Deltas, Vulnerability and Climate Change: Migration and Adaptation.

Sarkar, Sutapa Chatterjee. 2010. *The Sundarbans: Folk Deities, Monsters and Mortals*. New Delhi: Social Science Press.

Sarkhel, Prasenjit. 2015. "Flood Risk, Land Use and Private Participation in Embankment Maintenance in Indian Sundarbans." *Ecological Economics* 118:272–284.

Schendel, Willem van, and Aminul Haque Faraizi. 1984. *Rural Labourers in Bengal, 1880 to 1980*. Rotterdam: Comparative Asian Studies Program, Erasmus University Rotterdam.

Schlehe, Judith. 2010. "Anthropology of Religion: Disasters and the Representations of Tradition and Modernity." *Religions* 40 (2): 112–120.

Schmuck-Widmann, Hanna. 2001. *Facing the Jamuna River: Indigenous and Engineering Knowledge in Bangladesh*. Dhaka: Bangladesh Resource Center for Indigenous Knowledge.

Scott, James C. 1998. *Seeing Like a State: How Certain Schemes to Improve the Human Condition Have Failed*. New Haven, CT: Yale University Press.

Sen, Amartya. 1981. *Poverty and Famines: An Essay on Entitlement and Deprivation*. New Delhi: Oxford University Press.

Sen, H. S., ed. 2019. *The Sundarbans: A Disaster-Prone Eco-Region: Increasing Livelihood Security.* Cham: Springer.

Sen, Prasad. 2008. "River Bank Erosion and Protection in Gangetic Delta." In *River Bank Erosion and Land Loss*, edited by Kazi M. B. Rahim, Malay Mukhopadhyay, and Debashis Sarkar, 1–31. Kolkata: Visva-Bharati.

Sengupta, Debjani. 2011. "From Dandakaranya to Marichjhapi: Rehabilitation, Representation and the Partition of Bengal (1947)." *Social Semiotics* 21 (1): 101–123.

Shah, Alpa. 2011. *In the Shadow of the State: Indigenous Politics, Environmentalism, and Insurgency in Jharkhand, India.* New Delhi: Oxford University Press.

Sharma, Mukul. 2017. *Caste and Nature: Dalits and Indian Environmental Policies.* New Delhi: Oxford University Press.

Sharma, Sanjay. 2001. *Famine, Philanthropy, and the Colonial State: North India in the Early Nineteenth Century.* New Delhi: Oxford University Press.

Simone, AbdouMaliq. 2004. "People as Infrastructure: Intersecting Fragments in Johannesburg." *Public Culture* 16 (3): 407–429.

Simpson, Edward. 2011. "Blame Narratives and Religious Reason in the Aftermath of the 2001 Gujarat Earthquake." *South Asia: Journal of South Asia Studies* 34 (3): 421–438.

———. 2014. *The Political Biography of an Earthquake: Aftermath and Amnesia in Gujarat, India.* London: Hurst.

Simpson, Edward, and Malathi de Alwia. 2008. "Remembering Natural Disaster: Politics and Culture of Memorials in Gujarat and Sri Lanka." *Anthropology Today* 24 (4): 3–12.

Sinha, Indra. 2008. *Animal's People: A Novel.* New York: Simon and Schuster.

Sinha, Nitin. 2014. "Fluvial Landscape and the State: Property and the Gangetic Diaras in Colonial India, 1790s–1890s." *Environment and History* 20 (2): 209–237.

Sökefeld, Martin. 2012. "The Attabad Landslide and the Politics of Disaster in Gojal, Gilgit-Baltistan." In *Negotiating Disasters: Politics, Representation, Meanings*, edited by Ute Luig, 175–204. Frankfurt: Peter Lang.

Spalding, Mark, Anna Mcivor, Femke Tonneijck, Susanna Tol, and Pieter van Eijk. 2014. *Mangroves for Coastal Defence: Guidelines for Coastal Managers and Policy Makers.* Wageningen: Wetlands International and the Nature Conservancy.

Spencer, Jonathan. 2007. *Anthropology, Politics, and the State: Democracy and Violence in South Asia.* Cambridge: Cambridge University Press.

Spivak, Gayatri Chakravorty. 1988. "Can the Subaltern Speak?" In *Marxism and the Interpretation of Culture,* edited by Nelson Cary and Lawrence Grossberg, 271–313. Urbana: University of Illinois Press.

Statesman News Service. 2021. "People Pin Hopes of Mangrove Saving Them from Cyclone Yass's Fury." Bhubaneshwar, *The Statesman,* May 26, 2021. https://www.thestatesman.com/cities/bhubaneshwar/people-pin-hopes-of -mangrove-saving-them-from-cyclone-yasss-fury-1502968982.html.

Stoler, Ann Laura. 2008. "Imperial Debris: Reflections on Ruins and Ruination." *Cultural Anthropology* 23 (2): 191–219.

———, ed. 2013. *Imperial Debris: On Ruins and Ruination.* Durham, NC: Duke University Press.

Strang, Veronica. 2004. *The Meaning of Water.* Oxford: Berg.

Suhrke, Astri. 1994. "Environmental Degradation and Population Flows." *Journal of International Affairs* 47 (2): 473–496.

Sultana, Farhana. 2010. "Living in Hazardous Waterscapes: Gendered Vulnerabilities and Experiences of Floods and Disasters." *Environmental Hazards* 9 (1): 43–53. https://doi.org/10.3763/ehaz.2010.SI02.

Sundarbanbasir Sathe. 2010. *Kamon Ache Sundarbander Manus: Aila Parbarti Ekti Samiksa* [How are the people of the Sundarbans? A review of the time after Aila]. Kolkata: Manthan Samayiki.

Tarlo, Emma. 2003. *Unsettling Memories: Narratives of the Emergency in Delhi.* New Delhi: Permanent Black.

Tauger, Mark B. 2003. "Entitlement, Shortage and the 1943 Bengal Famine: Another Look." *Journal of Peasant Studies* 31 (1): 45–72.

Thiranagama, Sharika. 2007. "Moving On? Generating Homes in the Future for Displaced Northern Muslims in Sri Lanka." In *Ghosts of Memory: Essays on Remembrance and Relatedness,* edited by Janet Carsten, 126–149. Malden, MA: Blackwell Publishers.

Tironi, Manuel. 2018. "Hypo-Interventions: Intimate Activism in Toxic Environments." *Social Studies of Science* 48 (3): 438–455.

Tsing, Anna Lowenhaupt. 2015. *The Mushroom at the End of the World: On the Possibility of Life in Capitalist Ruins.* Princeton, NJ: Princeton University Press.

Tsing, Anna Lowenhaupt, Heather Anne Swanson, Elaine Gan, and Nils Bubandt, eds. 2017. *Arts of Living on a Damaged Planet: Ghosts and Monsters of the Anthropocene.* Minneapolis: University of Minnesota Press.

Tweed, Thomas A. 2006. *Crossing and Dwelling: A Theory of Religion.* Cambridge, MA: Harvard University Press.

Ullberg, Susann. 2010. "Disaster Memoryscapes." *Anthropology News* 51 (7): 12–15.

Unnikrishnan, A. S., and D. Shankar. 2007. "Are Sea-Level-Rise Trends along the Coasts of the North Indian Ocean Consistent with Global Estimates?" *Global and Planetary Change* 57:301–307.

Urry, John. 2007. *Mobilities*. London: Polity.

Valmiki. 2005. *Ramayana Book One: Boyhood*. New York: New York University Press.

Veer, Peter van der. 1988. *Gods on Earth: The Management of Religious Experience and Identity in a North Indian Pilgrimage Centre*. London: Athlone Press.

Vigh, Henrik. 2008. "Crisis and Chronicity: Anthropological Perspectives on Continuous Conflict and Decline." *Ethnos* 73 (1): 5–24.

von Schnitzler, Antina. 2013. "Traveling Technologies: Infrastructure, Ethical Regimes, and the Materiality of Politics in South Africa." *Cultural Anthropology* 28 (4): 670–693.

Wadley, Susan S. 1983. "The Rains of Estrangement: Understanding the Hindu Yearly Cycle." *Contributions to Indian Sociology* 17 (1): 51–85.

Wakefield, Stephanie. 2019. "Forum 4: Amphibious Architecture beyond the Levee." *Mobilities* 14:1–7.

Wakefield, Stephanie, and Bruce Braun. 2018. "Oystertecture: Infrastructure, Profanation, and the Sacred Figure of the Human." In *Infrastructure, Environment, and Life in the Anthropocene*, edited by Kregg Hetherington, 193–215. Durham, NC: Duke University Press.

Watts, Michael J. 1983. *Silent Violence: Food, Famine and Peasantry in Northern Nigeria*. Berkeley: University of California Press.

Werbner, Pnina. 1980. "Rich Man Poor Man: Or a Community of Suffering: Heroic Motifs in Manchester Pakistani Life Histories." *Oral History* 8 (1): 43–48.

———. 1996. "The Fusion of Identities: Political Passion and the Cultural Performance among British Pakistanis." In *The Politics of Cultural Performance*, edited by David Parkin, Lionel Caplan, and Humphrey Fisher, 81–100. Oxford: Berghahn.

Wisner, Ben, Piers Blaikie, Terry Cannon, and Ian Davis. 2004. *At Risk: Natural Hazards, People's Vulnerability and Disasters*. New York: Routledge.

WWF-India. 2010. *Voices of Change: Stories from Ladakh and Sundarbans*. New Delhi: WWF-India. http://awsassets.wwfindia.org/downloads/voices_of _change.pdf.

Zaman, Mohammad Q. 1989. "The Social and Political Context of Adjustment to Riverbank Erosion Hazard and Population Resettlement in Bangladesh." *Human Organization* 48 (3): 196–205.

————. 1991. "Social Structure and Process in Char Land Settlement in the Brahmaputra-Jamuna Floodplain." *Man* 26 (4): 673–690.

————. 1999. "Vulnerability, Disaster, And Survival in Bangladesh: Three Case Studies." In *The Angry Earth: Disasters in Anthropological Perspective*, edited by Anthony Oliver-Smith and Susanna M. Hoffman, 162–178. New York: Routledge.

Index

abandonment, 144–145, 156–157; of outer ring embankments, 158–159
absence, 9, 19
abundance, 109–110. *See also* rural riches
accretion, 7, 9, 12, 21, 34
Adivasi, 88, 90
aftermath, 54–55
Aila bandh, 151
amphibious, 12
Anthropocene, 8, 31
anticipation, 84, 86–87
archipelago, 14
attuning, 6

Bangladesh, 14
Bankimnagar, 10, 127–131
beach tourism, 10, 40
belonging, 136, 141
bhangon, 30, 50, 61, 63–64, 89, 103, 111, 176, 181; making sense of, 174. *See also* coastal erosion
Bhatir desh, 35
blame, 33, 175–184, 197
block, 123–124
boat accidents, 50, 94, 189
bouldering, 155–156
brackish, 15
bride trafficking, 41

Calcutta, 37, 105. *See also* Kolkata
care, 19–20; and maintenance work, 166, 169; state care, 151, 200
caste, 10, 16–17, 37
char, 22–23, 70–71

circuits of displacement and emplacement, 4, 33, 89, 108, 115
claustrophobic certainty, 8, 32, 48, 65–66, 144, 165–167, 174
climate change, 30–31, 116–117, 183–184, 204; provincializing climate change, 31; Sundarbans as climate change hotspot, 28–29; as tsunami, 184
climate refugees, 16, 29, 40, 116–117
coastal erosion, 9, 11, 12, 45; as collective predicament, 81–82, 85; as individualizing experience, 78–79; making sense of, 174; temporal dimensions of, 8, 47. *See also* *bhangon*
coastal futures, 19, 204–205
coastal heterogeneity, 12, 19
coastal protection, 145–147, 200; impermanence of, 148–151, 156; mangroves and/as coastal protection, 163–164; minimalist forms of, 157–159; mosaic of protection measures, 19, 147, 152, 156. *See also* embankments
coastal resilience, 46
colonial settlement, 14, 35, 41, 90–92, 96, 99–102
Colony Para, 4, 10, 138–142
compaction, 62–63
confluence of river and sea, 176
CPI(M), 38, 58, 102. *See also* Left Front
currents, 11, 51, 63, 65, 154, 170, 174, 182

cyclone, 20, 94, 100, 105; Cyclone Aila, 48–49, 190; Cyclone Amphan, 49, 191; Cyclone Bulbul, 49, 191; Midnapure Cyclone (Red Flood), 50, 105; protection from, 191–192
cyclone shelter, 5, 44–47, 154

Dabhlat, 130–131
death, 66–69, 129
deafening certainty, 48, 66
delta, 7
delta geomorphology, 21, 35
desh, 136
dharma, 179
disaster, 47–54, 87, 105, 201; disaster preparedness, 45; state disaster management, 100, 106. See also humanitarian interventions
distributed disaster, 20, 28, 40, 56, 202; coastal erosion as distributed disaster, 47–57; as collective predicament, 77, 79, 81; as individualizing experience, 78; and spikes, 73, 84; temporalities, 60–61
dredging, 182

embankments, 3, 33; colonial management, 42; and environmental knowledge, 148; as instruments of landscape transformation, 147; and politics of maintenance, 147–148, 160–167, 169–170; postcolonial management, 42–43
endurance, 86, 114, 205
enigma of departure, 88–89
entrepreneurial selves, 107–108, 115
environmental displacement, 26, 70, 87; as dis-placement, 26, 85; and resettlement, 120–123; standard narrative of, 70, 77
environment and migration, 23–26; immobility, 24; migration as adaptation, 24–25, 118. See also labor migration
equality, 37, 107, 108

erosion, 3, 7–9; river-bank erosion, 21; soil erosion, 9. See also coastal erosion
eventfulness, 52, 69, 73, 84
evictions, 112

famine, 104
fish, 63, 109, 132; becoming fish, 51; as ritual offering, 189. See also meen
fishermen, 50, 111, 176, 185
fishing trawlers, 11, 172
floods, 35–36, 50–51, 63; floods and reordering of property relations, 51; saline water floods, 22; type of floods, 63–64. See also Midnapore Cyclone (Red Flood)
fluid divinity, 94, 174–175
fluidity, 22
forced migration, 24, 88, 104
forest frontier, 37, 87, 92, 98, 102, 108, 110, 121; and egalitarianism, 17
forgetting, 27–28
full moon, 57, 59–60
futures, 18–19, 30

Ganga, 176–180, 203; ambiguity toward, 177–178; as fluid divinity, 177; Ganga Puja, 187–189; the other Ganga, 178–179. See also pilgrimage
Gangasagar, 1, 4, 10
Gangasagar Colony, 1, 4. See also Colony Para
Gangasagar Mela, 10, 40, 95, 135, 188–189
gendered vulnerability, 50, 169–170
Ghoramara, 4, 9, 12, 29, 90–92
ghosts, 129–130
goddesses, 60. See also hot goddesses
Gram Panchayat, 121, 124, 156

Haldia, 39, 181
hazards, 16
high land, 65–66
Hindu cosmology, 92–94. See also sacred geography
Hindu nationalism, 37

horse, 89–93
hot goddesses, 185–186
houses, 74–76, 144–145; abandoning
     houses, 74–75; ruins as segments
     of coastal protection, 166–167,
     169
Hugli River, 38, 180
humanitarian interventions, 44–47,
     53, 83

impoverishment, 72–73
Indira Awas Yojana (IAY), 84, 121,
     139–140
industrial development, 39, 112
infanticide, 97–98
infrastructure, 134; lively infrastructure,
     147
infrastructure development, 133–135,
     195; and localized belonging,
     135–137
inheritance, 72, 101, 102, 126, 139
instruments of landscape transforma-
     tion, 147, 156
itinerant slums, 79–81

Jambudvip, 39, 111
jungle, 90–92, 97–98

Kapil Muni, 92–94, 191–193, 195; and
     masculine protectors, 193
King Sagar, 92–96, 191
Kolkata, 13, 37, 171. See also Calcutta

labor migration, 11, 25, 41
land as potentiality, 87
landlessness, 17
land reform, 42, 120–121
land scarcity, 41, 90–91, 140
land titles, 121, 125–126; and gender,
     126–127; as instable, 140–141
Left Front, 121, 132, 195
life histories, 88
lighthouse, 59, 150, 157
liquification of land, 5, 51
livelihood, 10, 176, 184
Lohachara, 4, 9
loss, 18–19, 26–27, 200

love islands, 107–108
low land, 61–63

Makar Sakranti, 188
Mamata Banerjee, 39, 151, 194–195
mangroves, 45, 99, 111, 162–164
Marichjhapi, 14, 38, 112
marine traffic, 97, 150–151, 154,
     181–183. See also lighthouse
massacre, 113. See also Marichjhapi
master plan, 82, 157
materiality, 5
media, 30, 40, 86, 116–117
meen, 11, 184
memory, 27; disaster and memory, 27;
     memory and forgetting, 27–28
micromigration, 12, 24, 25, 76–77, 199;
     and repeat displacement, 76; of
     deities, 194
middlemen, 104
minor infrastructures, 33, 148,
     166–170
mishti, 22
MNREGA, 160
mobile matter, 23
mobility, 23, 118
mohima, 179
monsoon, 5, 21, 35
Muslim, 10, 193–194

Nandigram, 38–39, 113
nature conservation, 15
Nayachar, 11, 39, 112–113

paddy, 168
pan (piper betle), 165, 168
paper titles, 84, 139
Partition, 38
party politics, 38, 58, 135, 138–139,
     153, 160
past futures, 28
pilgrimage, 10, 40, 93–94, 98–99, 177,
     189; as income stream, 131–133.
     See also Gangasagar Mela
pilgrimage center, 4, 94, 131–134, 137,
     177, 195
pirates, 99

Pirbaba, 193–194
planetary injury, 8, 17
planned retreat, 25, 119, 120
politics of uneven provisioning, 152
population density, 13, 19
port authorities, 149, 182. *See also*
marine traffic
protection, 173, 184, 191, 193
proxy witness, 30

Ramayana, 92–96
refugee imaginary, 86–88, 127
refugees, 38. *See also* climate refugees
relearning, 5–6, 28, 44
religion, 16
resettlement, 33, 117–118, 120; politics
of resettlement, 121–123; uneven
policies of, 83–84. *See also* unacknowledged relocation
resettlement colony, 4
resilience, 49, 163; coastal resilience,
46, 202
retreat, 3
ritual bathing, 185
river water, 11
ruination, 7
ruins, 5
rural riches, 109–110
rural well-being, 37, 198

sacred geography, 40, 94
safety, 58–59
Sagar, 10; as beyond state authority,
94–96, 107; as Kapil Muni's territory, 95, 103
sea level rise, 4, 16, 31, 120, 204; laboratory for living with sea level rise,
21, 28
settlement of swamps, 12, 109, 127

silence, 71, 77–78, 172–173; silencing
the shore's advance, 71
siltation, 7, 21
Singur, 113
slow violence, 26, 55, 84
Small Island Development States, 13
socialities, 7
Special Economy Zone, 112–113
spectator, 29, 69
spectrality, 166–167
spring tides, 59
storms, 42, 45. *See also* cyclones
Sundarbans, 9, 13–17; as climate
frontier, 16, 28–29; Sundarbans
UNESCO Biosphere Reserve, 36;
as unsuitable for human habitation
15, 100, 200. *See also* Bhatir desh

terrestrial thinking, 23, 42, 199
thinning, 7, 18–20, 23
tiger, 14, 16, 36, 40, 94, 98
tirtha, 95, 177
tourist development, 36, 40, 111,
132
trade unions, 134
trapped populations, 76, 87, 199
trees, 68

unacknowledged relocation, 122
uncertainty, 33, 173, 178, 183, 197
uneventfulness, 54–57. *See also*
eventfulness

violence, 101–102
vulnerability, 52, 103

water, 11; brackish water, 15; drinking
water, 6; unnatural waters, 6, 8, 34,
174, 183, 190, 196–197, 200

About the Author

ARNE HARMS is a senior research fellow at the Max Planck Institute for Social Anthropology, Halle/Saale. He is an environmental anthropologist working on environmental degradation and disasters, everyday activism, and ethics.